# Science Works

## 1

Philippa Gardom-Hulme

Pam Large

Sandra Mitchell

Chris Sherry

D0317248

OXFORD

UNIVERSITY PRESS

# OXFORD
## UNIVERSITY PRESS

Great Clarendon Street, Oxford OX2 6DP

Oxford University Press is a department of the University of Oxford.
It furthers the University's objective of excellence in research, scholarship,
and education by publishing worldwide in

Oxford   New York

Auckland   Cape Town   Dar es Salaam   Hong Kong   Karachi
Kuala Lumpur   Madrid   Melbourne   Mexico City   Nairobi
New Delhi   Shanghai   Taipei   Toronto

With offices in

Argentina   Austria   Brazil   Chile   Czech Republic   France   Greece
Guatemala   Hungary   Italy   Japan   Poland   Portugal   Singapore
South Korea   Switzerland   Thailand   Turkey   Ukraine   Vietnam

© Oxford University Press

British Library Cataloguing in Publication Data

Data available

ISBN-13: 9780-19-915245-2

10 9 8 7 6 5 4 3 2 1

Printed in Spain by Cayfosa

Paper used in the production of this book is a natural, recyclable product made
from wood grown in sustainable forests. The manufacturing process conforms to
the environmental regulations to the country of origin.

## Acknowledgments

The Publisher would like to thank the following for permission to reproduce
photographs: p6tc Andrew Syred/Science Photo Library; p6bc Laguna
Design/Science Photo Library; p6t Andrew Syred/Science Photo Library; p6b James
Antrim/iStockphoto; p7tc Scott Bauer/US Department of Agriculture/Science Photo
Library; p7bc Andrew Syred/Science Photo Library; p7t Eye of Science/Science
Photo Library; p8t Dr Gopal Murti/Science Photo Library; p8b Dr Jeremy Burgess/
Science Photo Library ; p9 J.C. Revy/Science Photo Library; p10 Eye of Science/
Science Photo Library; p12t Eye of Science/Science Photo Library; p12b Andrew
Syred/Science Photo Library; p13t Prof. P. Motta/Dept. of Anatomy/ University "La
Sapienza", Rome./Science Photo Library; p13c Steve Gschmeissner/ Science Photo
Library; p13b Dr Tony Brain/Science Photo Library; p14t Gustoimages/Science
Photo Library; p14c Prof. P. Motta/Dept. of Anatomy/ University "La Sapienza",
Rome./Science Photo Library; p14b akg-images; p15t Anatomical Travelogue/
Science Photo Library; p15c Herve Conge, ISM/Science Photo Library; p15b Visual
Arts Library (London)/Alamy; p16 AJ Photo/Science Photo Library; p17tl CNRI/
Science Photo Library; p17tr Ed Reschke/Peter Arnold Inc./Science Photo Library;
p17b Crispin Hughes/Photofusion Picture Library; p18bl Eye Of Science/Science
Photo Library; p18br Pascal Goetgheluck/Science Photo Library; p20 Tom Grill/
Corbis UK Ltd.; p21 J.c. Revy/Science Photo Library; p22 Dr G. Moscoso/Science
Photo Library; p24 Neil Bromhall/Science Photo Library; p25 Publiphoto Diffusion/
Science Photo Library; p26t Dan Kelleher/iStockphoto; p26b Malcolm Schuyl/Frank
Lane Picture Agency; p27bl Francoise Sauze/Science Photo Library; p27br Bryan
Sage/Ardea; p27t Images of Africa Photobank/Alamy; p28 Bubbles Photolibrary/
Alamy; p30 Tom Grill/Corbis UK Ltd.; p32 Gary Houlder/ Corbis UK Ltd.; p33t James
King-Holmes/Science Photo Library; p33b Martin Dohrn/Science Photo Library; p34
Ruth Jenkinson/MIDIRS/Science Photo Library; p36l Paul Goff/DK Images; p36r
iStockphoto; p38 Human Solutions GmbH; p40t Photolibrary Group; p40b Chris
Howes/Wild Places Photography/Alamy; p42 Alessandro Di Meo/Epa/Corbis UK Ltd.;
p43 Bruno Domingos/Reuters/Action Images; p44t Mark Carwardine/Nature Picture
Library; p44c Carolyn A. McKeone/Science Photo Library; p45 Tom McHugh/Science
Photo Library; p46bl CNRI/Science Photo Library;
p46br Gregory Dimijian/Science Photo Library; p46t Profimedia International
s.r.o./Alamy; p47 Doug Martin/Science Photo Library; p48t Christopher Swann/
Science Photo Library; p48b B. G Thomson/ Science Photo Library; p49t Ken M.
Highfill/Science Photo Library; p49c Carolyn A. McKeone/Science Photo Library;
p49b Brandon Cole/Nature Picture Library; p50t Michael Clutson/Science Photo
Library; p50b Jim Steinberg/Science Photo Library; p51bl Science Photo Library;
p51br Science Photo Library; p51t Andrew Syred/ Science Photo Library; p52 John
Reader/Science Photo Library; p53t PA Photos; p53c Mauricio Anton/Science Photo
Library; p53b Pascal Goetgheluck/Science Photo Library; p64 Vladimir Dinets; p68
Ina Peters/iStockphoto; p70t John Cleare Mountain Camera; p70b Photographers
Direct/Robert Down Photography; p71bl Martyn F. Chillmaid; p71bc Martyn F.
Chillmaid; p71br Martyn F. Chillmaid; p71t Rob Hill/iStockphoto; p72 Peter Griffin/
Alamy; p73tl Martyn F. Chillmaid; p73tr Martyn F. Chillmaid; p73b Charles
O'rear/Corbis UK Ltd.; p74t Profimedia International s.r.o./Alamy; p74c Martyn F.
Chillmaid; p74b Martyn F. Chillmaid; p76 James King-Holmes/Science Photo
Library; p77 Martyn F. Chillmaid; p78t Photographers Direct/Kai Gedeon
Photography; p78c JORGEN SCHYTTE/Still Pictures/Still Pictures; p78b JORGEN
SCHYTTE/Still Pictures/Still Pictures; p79c JORGEN SCHYTTE/Still Pictures/Still
Pictures; p79c JORGEN SCHYTTE/Still Pictures/ Still Pictures; p79b JORGEN
SCHYTTE/Still Pictures/Still Pictures; p80l Steven Vidler/Eurasia Press/Corbis UK
Ltd.; p80c Leslie Garland Picture Library/Alamy; p80r Jack Sullivan/Alamy; p81cl
Martyn F. Chillmaid; p81l Martyn F. Chillmaid; p81cr Martyn F. Chillmaid; p81r
Martyn F. Chillmaid; p82tl Frances Twitty/ iStockphoto; p82tr ACE STOCK
LIMITED/Alamy; p82c f1 online/Alamy; p82b Charles D. Winters/Science Photo
Library; p83bl Dan Møller/iStockphoto; p83bc Mark A. Schneider/Science Photo
Library; p83br Pasieka/Science Photo Library; p83t Leslie Garland Picture Library/
Alamy; p84t Axel Hess/Alamy; p84c Andrew Lambert Photography/Science Photo
Library; p84b sciencephotos/Alamy; p85l Martyn F. Chillmaid; p85c Martyn F.
Chillmaid; p85r Martyn F. Chillmaid; p86 Photographers Direct/Mark Lane
Photography; p87 Zooid Pictures; p88l Annabella Bluesky/Science Photo Library;
p88c Martyn F. Chillmaid; p89 Martina Meyer/ iStockphoto ; p90l bildagentur-
online.com/th-foto/Alamy ; p90c Rob Wilkinson/ Alamy ; p90r Magnus Hjorleifsson/
Nordicphotos/Alamy ; p92 Trip/Alamy ; p93 Photographers Direct/John Rattle ;
p94bl Martyn F. Chillmaid; p94br Martyn F. Chillmaid ; p94t David Lee/
iStockphoto ; p95tl Martyn F. Chillmaid ; p95tc Martyn F. Chillmaid ; p95bc
Martyn F. Chillmaid ; p95tr Martyn F. Chillmaid ; p95b Andrew Lambert
Photography/Science Photo Library ; p96 Foodpix/Photolibrary Group ; p97 Michael
St. Maur Sheil/Corbis UK Ltd. ; p98t Jack Sullivan/Alamy ; p98c Gay Bumgarner/
Alamy ; p98b Photographers Direct/Tim Scrivener Agricultural Photography ; p99t
GIPhotoStock/Alamy ; p99c SCPhotos/Alamy ; p99b ImageState/ Alamy; p100l Dan
Guravich/Corbis UK Ltd.; p100c Holt Studios International Ltd/Alamy; p100r Ashley
Cooper/Alamy; p102l Apple Computer Inc.; p102tc Jim DeLillo/iStockphoto; p102bc
iStockphoto; p102cr Matjaz Boncina/ iStockphoto; p102r Deborah Chadbourne/
Alamy; p102b Artem Efimov/ iStockphoto; p106l Bradley Mason/iStockphoto;
p106r Ivan Stevanovic/iStockphoto; p108tl Jasmin Awad/iStockphoto; p108tr
Andrew Lambert Photography/Science Photo Library; p108b Thomas Mounsey/
iStockphoto; p109 Thomas Mounsey/ iStockphoto; p111 Kevin Foy/Alamy; p112t
Danita Delimont/Alamy; p112c akg-images; p112b akg-images; p113 Mary Evans
Picture Library/Alamy; p114tl Bob Thomas/iStockphoto; p114bl Dane Wirtzfeld/
iStockphoto; p114tc Artur Achtelik/ iStockphoto; p114tr Randy Mayes/iStockphoto;
p114br Alexandra Draghici/ iStockphoto; p115tl iStockphoto; p115tr Jake Hallman/
iStockphoto; p115b Eric Hood/iStockphoto; p116t Zooid Pictures; p116b Zooid
Pictures; p117l Andrew Ramsay/iStockphoto; p117c ACE STOCK LIMITED/Alamy;
p117r Jeannette Meier Kamer/iStockphoto; p118tl Clint Spencer/iStockphoto;
p118bl Clint Spencer/ iStockphoto; p118bcl iStockphoto; p118tc Raoul Vernede/
iStockphoto; p118tr Isabel Massé/iStockphoto; p118br f1 online/Alamy; p118bcr
Willi Schmitz/ iStockphoto; p119 Webstream/ Alamy; p120t Ronald Bloom/
iStockphoto; p120c Scott Leigh/iStockphoto; p121 Kieran Mithani/iStockphoto;
p122t Timothy Hughes/iStockphoto; p122b Dave White/iStockphoto; p123t Todd
Arbini/ iStockphoto; p123c Tim McCaig/ iStockphoto; p123b Bryan Busovicki/
iStockphoto; p124t Andrew Johnson/ iStockphoto; p124b EDF; p125t iStockphoto;
p125b Ron Yue/Alamy; p126bl Pete Jenkins/Alamy; p126tc BE&W agencja
fotograficzna Sp. z o.o./Alamy; p126bc Bettmann/Corbis UK Ltd.; p126br Rick
Gomez/Corbis UK Ltd.; p126t Johnson Space Center/NASA; p127l Stephen Frink
Collection/Alamy; p127c ImageState/Alamy; p127r Andrew Lambert Photography/
Science Photo Library; p128t Mark Dadswell/ Getty Images; p128c Ryan Pierse/
Getty Images; p128b Maciej Noskowsk/ iStockphoto; p129 Colin C. Hill/Alamy;
p130t Matt Craven/iStockphoto; p130b Christian Carroll/iStockphoto; p131 BMW;
p132t Amy Walters/iStockphoto; p132b Doug Webb/Alamy; p133t Arco Images/
Alamy; p133c Martin Harvey/Alamy; p133b John Peter Photography/Alamy; p134t
National Motor Museum/MPL; p134b PA WIRE/PA Photos; p136tl Simone van den
Berg/iStockphoto; p136tc Trevor Know/ Alamy; p136bc Noel Toone; p136tr
Image100/Corbis UK Ltd.; p136c allOver photography/Alamy; p136b Medioimages/
Corbis UK Ltd.; p137bl Jim West/Alamy; p137br Robert Cocquyt/iStockphoto; p137t
Yvan Dubé/iStockphoto; p138t Danielle Pellegrini/Science Photo Library; p138b
NASA; p140tc NASA; p140bc NASA; p140tr NASA; p140b NASA; p141 NASA; p142
NASA; p143t Stefan Seip; p143b Dan Schechter/Science Photo Library; p144t NASA;
p144b EUMETSAT; p145t Raymonds Press Agency; p145b STS-116 Shuttle Crew/
NASA; p146t Anglo-Australian Observatory/Royal Observatory, Edinburgh/David
Malin Images; p146b Jason Ware; p147t Lockheed Martin; p148t Science Photo
Library; p148b Palomar Observatory/ California Institute of Technology; p149cl
NASA Jet Propulsion Laboratory (NASA-JPL)/NASA; p149cr NASA; p149t NASA;
p149b NASA

Cover image: Bananastock/Jupiter Images

Illustrations are by Rui Ricardo at Folio Art and technical illustrations are by
Barking Dog Art.

This book is called *Science Works* because it shows you how scientists work out their ideas about the world, and how science can be put to work in everyday life.

It will help you:
- Develop your understanding of scientific ideas
- Work out scientific ideas for yourself using results from investigations
- See how science is used in everyday life
- Think about how we can use science for the best

The book has two types of pages. Most are like this:

'Learn about' lists your objectives for the topic

'Brainache' boxes have curious questions on the topic (and their answers!)

'Red' headings are to make you think harder!

'Green' sections revise ideas from Key Stage 2

'How do we...' boxes tell you something about how scientists work out their ideas

'Amber' sections move on to new ideas

'Get this' boxes tell what you should know from the topic

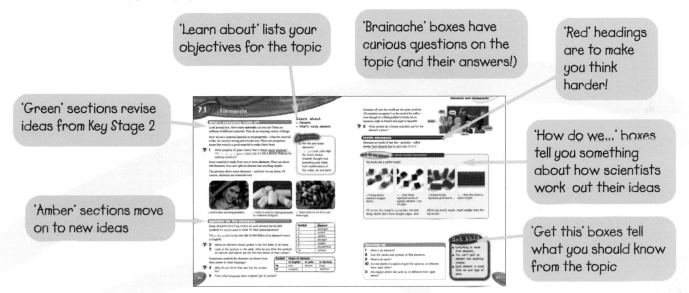

At the end of every unit there is a page called 'How Science Works' which looks at science in action. This might be how science is used in jobs or how science might affect our lives and how we can use it for the best.

All the key words you need to know are in **bold** and they are all defined in a glossary at the end of the book. You'll also find an index right at the back to help you find the information you want.

We hope *Science Works* will help you think about science, understand it and, above all, enjoy it.

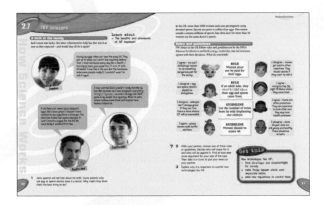

# Contents

shows which are the special 'How Science Works' pages (but you'll find out about how science works on other pages too).

Physics

### Feeling itchy?

They live in your hair and they suck your blood. But head lice are hard to spot. They're just too small.

A microscope can make them look bigger – it **magnifies** them. The picture you see under a microscope is called the **image**.

**1** Lice feed at least four times a day, and leave itchy red lumps on your scalp. Is there anything about their appearance that surprises you?

**2** These photographs were taken through a microscope. What has the microscope done to the images?

**3** The louse is really 2 mm long. How many times bigger does it look in image A?

**4** Image B has a higher magnification. How many times bigger than A is it?

### A whole new world

Magnifying glasses are **lenses**. You can fit them together to make powerful **microscopes**. The first were made in the 1600s, by Dutch lens makers. They were very popular.

One microscope maker wrote that people would 'cry out that they saw marvellous objects … a new theatre of nature, another world'. Ideas about life itself were starting to change.

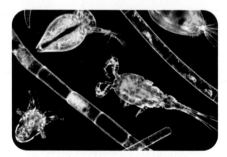

Tiny plants and animals like these are invisible without a microscope but they are the main food of many fish.

**5** How do you think microscopes started to change people's ideas about living things?

### Getting things in focus

School microscopes use **light** and **lenses** to magnify things.

**6** What do you call the thing you look at under a microscope?

**7** How do you get different magnifications?

**8** Why should you never point the mirror directly towards the Sun?

**9** The little blue louse is 1.5 mm long. How long would its image be if it was magnified 10 times?

**10** An image of the largest louse ever seen is 100 mm long and labelled (magnification ×20). How long was the specimen?

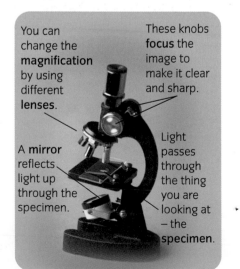

You can change the **magnification** by using different **lenses**.

These knobs **focus** the image to make it clear and sharp.

A **mirror** reflects light up through the specimen.

Light passes through the thing you are looking at – the **specimen**.

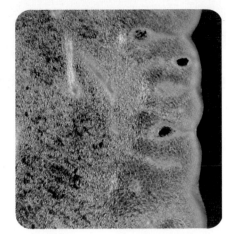

## Making things clear

Thin, see-through specimens give clear images in **light microscopes**. Thick specimens have to be sliced into thin **sections**.

If you could fit a lemon under a light microscope you'd just see a black blob. But this thin slice lets light through to make an image. It has been magnified 10 times.

**11** What is a section?

**12** Why do specimens need to be thin?

**13** How could you make sure you saw everything inside the lemon?

## Magnifying more times

Light microscopes magnify up to 1000 times. In universities and research labs, scientists use **electron microscopes**. They are very expensive, and need trained operators. But they give fantastic images – up to one million times the size of the specimen.

There are two sorts. SEMs (scanning electron microscopes) scan surfaces and TEMs look through thin sections. Images from SEMs are coloured artificially. This makes it easier to see the different parts of a specimen.

Things look different at high magnifications. This skin is magnified 500 times.

**14** What sort of electron microscope took the skin photograph?

**15** The skin is from the back of a young girl's hand. What can you see in the magnified image?

**16** Can we be sure that the skin was really this colour?

## Summing up

**17** What do light microscopes use to make things look bigger?

**18** Why do you usually need to take a thin slice of a specimen to look at it under a light microscope?

**19** Most specimens are dead when you see them under a light microscope. Explain why.

**20** Explain **two** things electron microscopes can show that light microscopes can't.

**21** What type of microscope do you think was used to get the images of the headlice opposite?

## Get this

- Microscopes let us see:
  - very tiny things
  - extra detail.
- Microscopes changed the way people thought about the world.

### Open wide

**Learn about**
- Cells
- Differences between animal and plant cells

The image below on the right shows skin from the insides of your cheeks. It has been magnified 2000 times.

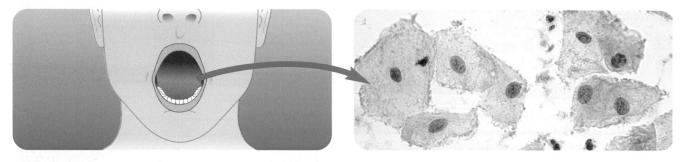

The tiny compartments are cells. Every bit of you is made of cells. It's the same for every living thing. Cells are so small that up to 50 million, million fit inside an adult human. That's 50 000 000 000 000 cells!!

### How do we tell others... what we see down a microscope?

Early scientists had to draw and describe what they saw. Cells get their name from an early description. In 1665, Robert Hooke published this microscope drawing of the cork used for wine bottle stoppers. 'It is full of tiny boxes like the bare rooms monks use' he added. The monk's rooms were called **cells** and the name stuck. Drawing and describing are still important skills.

**1** The cells inside your cheeks are stuck together. They separate when they are scraped out. Draw the outline of a single cell.

**2** In a different colour, add any parts that Hooke didn't draw.

**3** Now describe the differences in words. Which is easier?

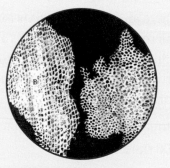

▲ Magnification ×50

**4** What extra information should you give with your drawing? Why might the photo show more detail than Hooke's drawing?

### Cells alive

The most important thing your cells do is *stay alive*, and that keeps you alive. Cells are the smallest units of life.

**5** Hooke's cells were dead and empty. Live cells have a nucleus, cytoplasm and cell membrane. Label these on your drawing.

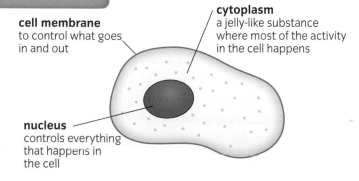

**cell membrane**
to control what goes in and out

**cytoplasm**
a jelly-like substance where most of the activity in the cell happens

**nucleus**
controls everything that happens in the cell

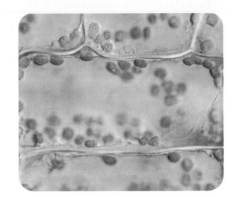

## Are plants the same?

Plant cells have a cell membrane, a nucleus and cytoplasm, like animal cells. But there are important differences:

 **6** The leaf cell has been magnified 2500 times. What can you see in it that cheek cells don't have?

Plants make food using **water**, **carbon dioxide** and **sunlight**. The green things are **chloroplasts**. They capture the energy from light.

The food is stored in **cell sap**, which fills a large bag called a **vacuole** in the middle of the cell.

Plants cells have walls around their membranes made of tough **cellulose** fibres. Vacuoles push against the **cell walls** to keep plant cells firm.

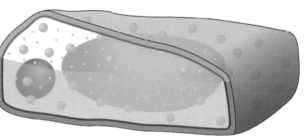

Light microscopes can make cells look flat and green all through, but they aren't really. This diagram shows what a plant cell is really like.

Flat, 2D diagrams can show a cell's features more clearly than a photograph and are easier to draw.

**7** Below is a 2D diagram of a plant cell. Copy the diagram and put the correct labels on parts A–F.

**8** How do the chloroplasts help a plant to make its food?

**9** List **two** uses of vacuoles.

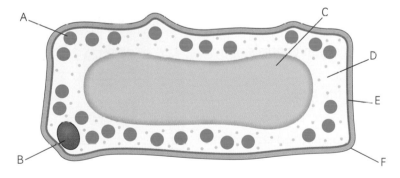

## Summing up

**10** What **three** parts do all cells have?

**11** Name the **three** extra features most plant cells have.

**12** Plant root cells do not have chloroplasts. Why?

**13** When plant cells are short of water their vacuoles get smaller. The cells go floppy and the whole plant droops. Explain why.

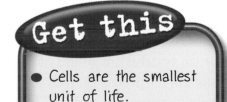

### Get this

- Cells are the smallest unit of life.
- All cells have a nucleus, cell membrane and cytoplasm.
- Plant cells have extra structures for support and making food.

# Staying alive

## Cells alone

If you look at mud from a pond under a microscope, you may find one of these. It's not a plant or an animal. It's an amoeba. It looks like a blob of jelly but its shape keeps changing. It only has one cell to carry out all its life processes.

**1** How can you tell the amoeba is not a plant?

**2** People used to think the amoeba was an animal. What makes it look like an animal cell?

The amoeba **moves** by stretching out sections of cytoplasm and pulling the rest of the cell along behind. If it **senses** food, the cytoplasm flows out to surround it. The dark specks are trapped food. The amoeba breaks them down to get **nutrients**.

**3** Which **two** processes rely on an amoeba's ability to change the shape of its cytoplasm?

## Well organised

The amoeba has to do everything with one cell. We have billions of cells but they don't all do the same thing. They are organised into organs. Each organ specialises in a different life process.

Some of your organs are huge and others are tiny. Your skin is your biggest organ. It covers and protects all the others.

**4** The amoeba's nucleus controls its cell. Which organ controls our bodies?

**5** Name the organ which pumps blood around the body.

**6** What is organ C and what does it do?

**7** Which label shows your stomach, where food starts to be digested?

## Part of a system

**Organ systems** are groups of organs. They work together to carry out different life processes. Your biggest organ system is your **skeleto–muscular** system. It contains all the bones in your skeleton and the muscles that move them. But you wouldn't be able to move if you didn't have your nervous, circulatory and digestive systems too.

**8** Which system lets us sense where we are?

**9** Name the system that gets nutrients from our food?

**10** Which system carries nutrients to our muscles?

Learn about
- Life processes
- Cells, tissues and organs

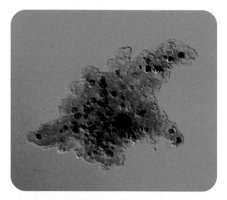

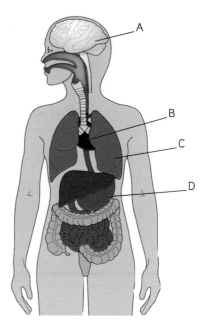

| Nervous system | Circulatory system | Digestive system |
|---|---|---|
| | | |
| **Organs** | **Organs** | **Organs** |
| brain, spinal cord, nerves, eyes, ears, nose, tongue and skin | heart, blood, blood vessels | mouth, stomach, intestines |

## The right tissue

Organs aren't the same all the way through. They contain different **tissues**. You have more than 200 different tissues and each does a different job. Most tissues can be found in more than one organ.

Tissues do different jobs because they contain different cells. All the cells in a tissue are the same. They are all specialised to do that tissue's job.

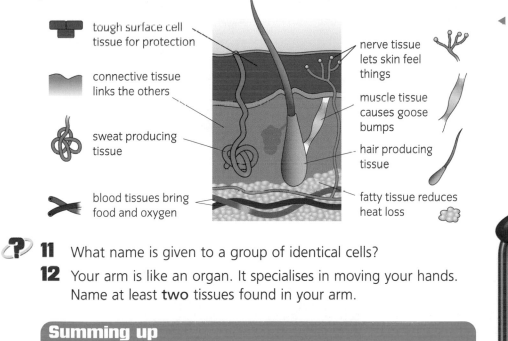

tough surface cell tissue for protection

connective tissue links the others

sweat producing tissue

blood tissues bring food and oxygen

nerve tissue lets skin feel things

muscle tissue causes goose bumps

hair producing tissue

fatty tissue reduces heat loss

◄ Skin – your biggest organ

**11** What name is given to a group of identical cells?

**12** Your arm is like an organ. It specialises in moving your hands. Name at least **two** tissues found in your arm.

## Summing up

**13** Put these body parts in order of size: tissue, organ system, cell, organ.

**14** How can you tell whether something is a tissue or an organ?

**15** Why do our organs contain so many different tissues?

## Get this

- Organs and organ systems carry out life processes.
- Organs have tissues that suit the jobs they do.
- Each tissue is made of a different type of cell.

# Right for the job

The cells that make up each tissue have special features which help them do their jobs.

## Cells that move

Forty percent of your tissues are **muscles** – less if you are female. Muscles work by **contracting**, which means getting shorter and fatter. As muscles shorten they pull on bones, and that makes you move.

Muscles work in pairs. As one muscle shortens and pulls, its partner relaxes and stretches. Bend your arm from the elbow. Which muscle contracts and which relaxes as your arm moves?

Muscle cells are very long and thin. The image below right shows short sections from four cells. They are packed into bunches so they can all pull together.

**?** **1** What sort of tissue joins muscles to bones?

**2** Which life process are muscle cells specialised for?

Your heart is mostly muscle. It pumps blood around your body for 24 hours a day.

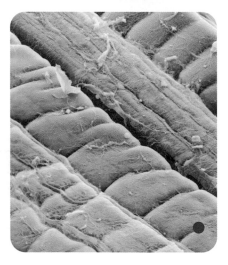

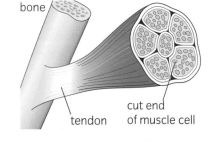

bone

tendon

cut end of muscle cell

## Cells that get pushed around

All cells use oxygen to get energy from food. They take the oxygen from your blood.

**?** **3** What makes blood travel around your body?

**4** Where does the blood get oxygen from?

Blood looks like a liquid, but it's full of cells – mostly red cells. More than half the cells in your body are **red blood cells**. They're tiny. These have been magnified 4000 times. All an adult's red cells would fit in a 2 litre coke bottle.

The red circle ● next to other cells on this spread shows how big red blood cells are compared to other cells.

Red blood cells take oxygen from your lungs to every cell in your body. They are packed with **haemoglobin** – a red chemical that carries oxygen.

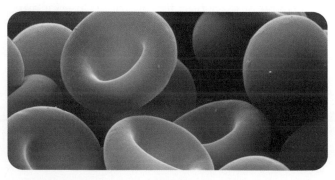

**5** What gives red cells their colour?

## Store cupboards

If you eat too much these huge pink cells get filled with oil. Around 25% of your body weight is fat cells – less if you are male. They store energy and stop you losing heat.

 **6** The oil in a fat cell squeezes its nucleus and cytoplasm up against its membrane. Draw the inside of a fat cell.

**7** People who lose weight often feel cold. Explain why?

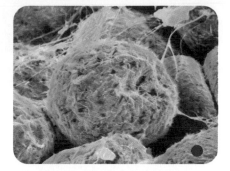

## Builders

Bones need to be rigid and hard to hold your body in shape. This hardness comes from minerals like calcium. Bone cells make fibres that minerals can stick to. The minerals turn the fibres into solid bone and the cells are trapped inside. Tiny holes in the bone let them contact other cells and pick up nutrients. Fifteen percent of your body is bone.

 **8** What are bones made of?

**9** How do we know that bones are alive?

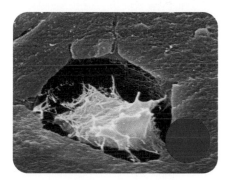

## Protectors

Surface cells protect your insides. Most are thin and flat and wear away quickly. The cheek cells on page 8 last a couple of days and the skin cells on page 7, a couple of weeks.

The cells that line your lungs are called **ciliated cells**. They are covered with tiny, hair-like cilia. These wave in unison to sweep dirt from the lungs. You find them wherever things need to be swept along tubes.

**10** Draw **one** cell and label the nucleus and cilia.

**11** Cigarette smoke stops the cilia moving. Why is that a problem?

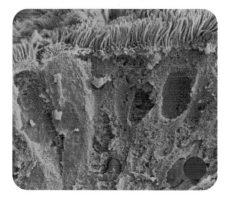

## Summing up

**12** Which **three** tissues make up 80% of your body?

**13** Muscles, ciliated cells and red cells all move. What is different about the way they do it?

**14** The images on this page are magnified by different amounts. Put them in order of size. Use the red cells as a guide.

**15** More than half the cells you have are red blood cells. Why can your blood carry so many cells?

**16** Cells need energy to carry out their jobs. Where do they get it from?

### Get this

- Cells are specialised for the jobs they do in the body.

# Extreme cells

### Wired for movement

Aibo acts like a dog but has motors instead of muscles and a computer for a brain. Wires link his computer to his motors so it can make him walk, lie down and get up again.

Nerves are your body's wires, and the cells inside them are extremely long and thin. They stretch out from your brain and spine to tell muscles when to move.

muscle cells

connections to nerves in your spine

**1** What controls Aibo's motors?

**2** What makes your muscles work when you decide to pick something up?

The fastest nerves send signals at 100 metres per second. This slice through some nerve cells has been magnified 350 times. They link the spine to a leg muscle, so they are about a metre long.

**3** What's unusual about the shape of a nerve cell?

**4** How tall would the photograph need to be to show the whole nerve cells magnified 350 times?

**5** If you damage nerve cells in your spine, you can't move your legs. Explain why.

**6** Estimate the length of the nerve cells that run from your spine to the bottom of your thumb.

### How do we know... about nerves

Ibn Sina (981 CE) wrote a book about medicine which was used in Europe until the 1600s. He said: nerves connect with the brain, make muscles move and pick up pain messages. So patients should be awake for some operations. Their screams show when the knife is cutting a live nerve – which proves their dead tissues have been successfully removed.

**7** How do you know when you have damaged a muscle?

**8** Why might a doctor ask a crash victim to wiggle her toes?

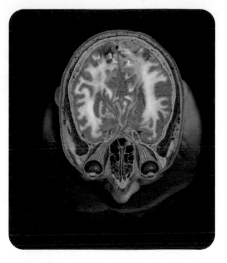

## Getting the message

Aibo can't feel pain but he can sense other things. The camera at the front of his head feeds data to his computer brain. The computer works out what he's looking at – a face or his coloured ball.

 **9** When Aibo's ball rolls in front of him, he walks towards it. What makes his computer turn his leg motors on?

Your eyes are like Aibo's camera. They send nerve impulses to your brain. Your brain works out what your eyes are detecting. It's the same for all your other senses.

**10** This picture shows the nerves from the eyes entering the brain. Can you see where their signals are processed?

## How do we know... that our brains work out what we can see?

Psychologists asked volunteers to wear special glasses that made everything look upside down. After a couple of days everything started to look the right way up again – until they took the glasses off.

**11** Why do you think the world started to look the right way up after a while?

**12** Why did it look the wrong way up again when the glasses were taken off?

## Making sense

Your brain has 100 billion nerve cells and they each connect to thousands of others. They process data from your senses so you can detect the world around you. This image shows a few of them.

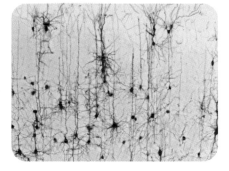

Making sense of data is difficult. The short cuts your brain takes make it easy to create illusions.

 **13** Archimboldo painted portraits in the 1500s. He relied on fooling our senses. Look at this image upside down and explain what happens.

## Summing up

**14** People used to think every part of our bodies was made of cells except the nerves and brain. Why were nerve cells so difficult to see under the microscope?

**15** Explain how nerves are involved when you take off some uncomfortable shoes.

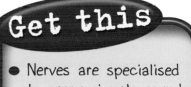

## Get this

- Nerves are specialised to carry signals round the body.
- Our brains work out what is happening around us.

### Going for tests

Sasha looks pale and never seems to have any energy. Her doctor thinks she might be **anaemic**. He has sent her for a blood test. If you're anaemic your blood can't carry enough oxygen and this makes you feel tired all the time. The nurse at the hospital is taking a blood sample to send to the lab.

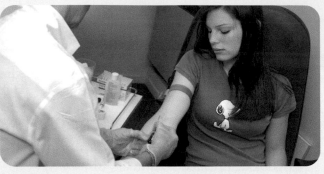

**1** How does blood carry oxygen?

**2** What could reduce the amount it can carry?

Biomedical scientists in the Haematology lab test more than 500 blood samples every day. Sasha's blood is put in a machine called an electronic blood-cell counter. The counter sucks up a small amount of her blood. Each cell in the blood is counted and measured as it moves through the machine.

**3** Sasha's blood cells could be put under a microscope and counted. Why is a machine used instead?

### Does Sasha have enough red cells?

The counter works out how many cells there are in each litre of Sasha's blood. She has 4 litres altogether and her red cell count is 3 billion cells per litre.

**4** Is her red cell count normal?

**5** Do all healthy people have the same red cell counts?

| Red cell count (billion per litre) | |
| --- | --- |
| Normal ranges: | |
| men | women |
| 4.5–5.6 | 3.8–5.0 |

### Do Sasha's cells have enough haemoglobin?

The machine also finds out how much haemoglobin there is in each decilitre of blood. Sasha's blood has 9.3 grams of haemoglobin per decilitre.

**6** Is her result normal?

| Haemoglobin (grams per decilitre) | |
| --- | --- |
| Normal ranges: | |
| men | women |
| 13–17 | 11–15 |

### Are her red cells the right size?

Sasha's results lie outside the normal range, so her blood is checked using a microscope. A drop of her blood is spread very thinly to make a blood smear. Then the red cells are compared with cells from a normal blood sample.

### Learn about
- Using microscopes to diagnose illness
- What red blood cell counts can tell us

1 litre

1 decilitre

**7** How much magnification is needed to show red cells clearly?

**8** Explain why the blood has to be spread very thinly.

**9** What do you notice about the colour of Sasha's red cells?

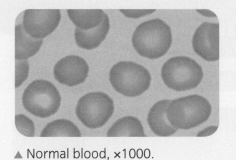

▲ Normal blood, ×1000.

▲ Sasha's blood, ×1000.

**10** Measure 5 cells from each blood sample. Add the measurements from each slide. Then divide by 5 to get average values.

**11** How do Sasha's cells compare with a normal sample?

## Taking a short cut

A quicker way to see if Sasha is anaemic is to measure how much of her blood is made up of red cells. The red cells are separated out so that they pack together at the bottom of a tube. The less space they take up the fewer or smaller the red cells are.

| Packed cell volume (%) | |
|---|---|
| Normal ranges: | |
| **men** | **women** |
| 38–50 | 36–46 |

Sasha's sample

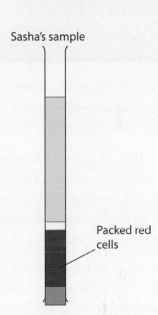

Packed red cells

**12** Measure the height of the red cells in Sacha's sample and the total height of the blood.

**13** Use this equation to find Sasha's result. Is it normal?

$$\text{packed cell volume \%} = \frac{\text{height of red cells}}{\text{total height}} \times 100\%$$

## Getting advice

Sasha's test results are sent to her doctor. They confirm that she is anaemic. The doctor explains what she needs to do.

'Your body needs iron to make haemoglobin. It looks as if you don't have enough in your diet. Try eating more eggs, red meat and dairy products, and take these iron tablets twice a day. If you don't start feeling better, come back. We can do more tests.'

**14** How can the doctor tell that Sasha is anaemic?

**15** Why does Sasha's doctor want her to eat more red meat?

**Get this**

- Biomedical scientists analyse tissue samples.
- Automatic machines make their jobs easier.

# Having a baby

## A new life

**Learn about**
- Egg and sperm cells
- Fertilisation
- Female sex organs

Jack's life began like everybody else's. A **sperm cell** from his dad **fertilised** one of his mother's **egg cells**. It's the same for most animals. Joining an egg and sperm to start a new life is called **fertilisation**.

This picture shows a scanning electron microscope image of two sex cells – an egg and a sperm (with added colour). A real egg cell is about the size of this full stop.

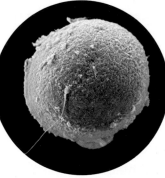

The other picture shows a fertilised egg seen through a light microscope. Extra sperm are stuck in the layer of jelly around the egg.

**1** Name the cells that join to make the first cell of a new baby.

**2** Write a sentence to describe what happens during fertilisation.

**3** Scanning electron microscopes show the outsides of cells. What extra detail can you see with a light microscope?

## How do we know... about fertilisation?

In 1786, Lazzaro Spallanzani put little pairs of shorts on male frogs. The frogs mated with females, but no tadpoles grew. So he scraped the sperm out of their shorts and put them on the eggs. This time he got loads of tadpoles.

**4** Why did the shorts stop the frogs making tadpoles?

**5** In the 1700s sperm were thought to give off a vapour that could fertilise eggs. Did Spallanzani's results support that idea?

## The egg's journey

A woman's eggs are all formed before she is born. They are stored in her **ovaries**. One egg usually matures each month. It bursts out of an ovary and gets swept along her oviduct. Egg cells can't move, but the cells which line the **oviduct** are coated with tiny cilia. These look like waving hairs and they sweep the egg along.

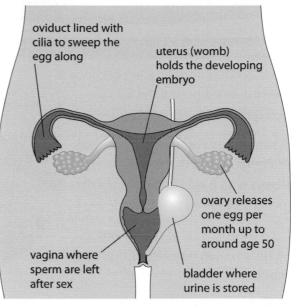

oviduct lined with cilia to sweep the egg along

uterus (womb) holds the developing embryo

ovary releases one egg per month up to around age 50

vagina where sperm are left after sex

bladder where urine is stored

**6** Which organs store egg cells until they mature?

**7** How often are egg cells usually released?

**8** How do egg cells get to the **uterus**?

When a couple have sex, sperm enter the woman's **vagina**. They can only survive for a day or two and they have a long swim ahead. They need to get to the oviduct to fertilise the egg. If fertilisation happens, an **embryo** forms, settles in the uterus and grows. Nine months later a baby is born.

**9** Where are sperm cells left during sex?

**10** Where does fertilisation usually happen?

**11** What is a baby called when it first starts to grow?

## Whose baby?

Jack's mum is lucky to be alive. She survived cancer. But the treatment that saved her life made it impossible for her to get pregnant. She couldn't have Jack without her sisters' help. Jack's mum's egg cells were damaged in her cancer treatment. Jack's egg came from his Aunt Jessica – his mum's sister. Jack was already an embryo when he was put into his mum's body.

Around 2000 babies in the UK are born using egg donation like this each year.

### Brainache

**Q** Why are oviducts sometimes called Fallopian tubes?

**A** Because an Italian professor called Gabriel Fallopius wrote the first clear description of them in the 1500s.

### Get this

● A new life begins when a sperm fertilises an egg.
● Eggs:
  ● mature in the ovaries
  ● are fertilised in the oviduct
  ● develop in the uterus.

## Summing up

**12** Write the story of an egg that is fertilised. Describe its journey from the ovary to becoming an embryo in the uterus.

**13** An egg gives off chemicals and these can attract a lot of sperm. How many sperm are needed to fertilise each egg?

**14** Many couples can't have a child without help. Jack's mum had faulty eggs. Think of **two** other problems that could make it hard to get pregnant.

# Becoming a dad

**Learn about**
- Male sex organs
- Having sex
- IVF

## Making sperm

Jack's dad has been making millions of sperm every day since he was a teenager. They form in his **testes**.

When he has sex, they get pumped along his **sperm ducts**. They pass **glands** which add fluids to help the sperm stay alive and swim.

Then the mixture of sperm and fluid – called **semen** – spurts out through his **penis**.

Ripe sperm will only keep for a fortnight. Any he doesn't use get recycled inside his body.

**?**
**1** Where are sperm made?
**2** Why are new sperm made all the time?
**3** What happens to sperm that aren't used?

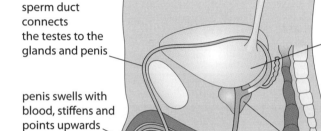

sperm duct connects the testes to the glands and penis

penis swells with blood, stiffens and points upwards during an erection

testes where sperm are made

bladder cannot empty when the penis is erect

glands add fluids that will keep the sperm alive

## Making love

Making love is a very intimate act, and it feels good. Couples usually kiss, stroke or hug each other to begin with.

The woman's vagina becomes moist and the man's penis swells with blood and stiffens. He has an **erection**. His erect penis can slip into his partner's vagina. Moving it forwards and backwards makes the pleasure more intense, he **ejaculates** and semen spurts into the vagina.

**?**
**4** What happens to a man's penis when he has an erection?
**5** What happens when a man ejaculates?
**6** What does semen contain?

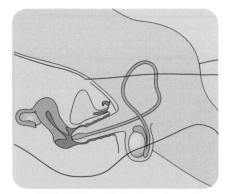

▲ A man's penis can slip into his partner's vagina when it is erect.

## IVF

Jack's embryo was made by **IVF** – *in vitro* fertilisation. A doctor fertilised 12 of Jessica's eggs with sperm from Jack's dad. Many embryos don't grow properly, but two good ones were placed in Jack's mum's womb. One died, sadly, but Jack survived. He was lucky. IVF only works the first time for 1 in 4 couples.

**7** When couples use IVF, where do the eggs get fertilised?

**8** Jack's parents only wanted one child. Why do you think the doctor took more than one egg?

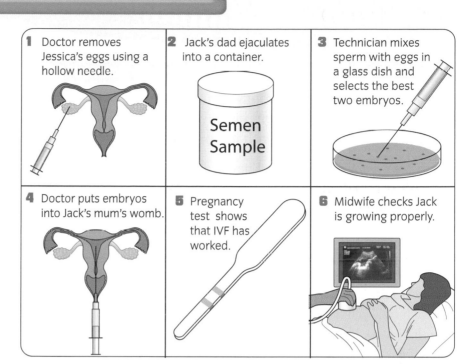

**1** Doctor removes Jessica's eggs using a hollow needle.

**2** Jack's dad ejaculates into a container.

Semen Sample

**3** Technician mixes sperm with eggs in a glass dish and selects the best two embryos.

**4** Doctor puts embryos into Jack's mum's womb.

**5** Pregnancy test shows that IVF has worked.

**6** Midwife checks Jack is growing properly.

## Babes on ice

In science fiction, passengers can be frozen for long trips through space. It can't be done in real life. But eggs, sperm and day-old embryos can be frozen and brought back to life.

The maximum legal storage time is 10 years – but they last much longer if kept frozen. Sperm are kept at −196 °C in special vacuum flasks. Each plastic straw in the picture contains a different person's sperm.

**9** Clinics that store frozen cells have alarms on their freezers and staff on call 24/7. Why is this important?

## Get this

- Sperm are:
  - made in the testes
  - pumped out of the penis
  - able to swim to the egg
  - often frozen for later use.
- In IVF eggs are fertilised in a dish.

## Summing up

**10** Describe a sperm's journey from the testes to the egg.

**11** Dev's sperm are poor swimmers. His wife cannot get pregnant. How could IVF help them?

**12** Ryan is very ill. The medicines he needs could stop his testes making sperm. What advice would you give him?

### The mating game

Your life began when an egg from your mother met a sperm from your father. This usually happens inside your mother's body, not in a test tube, and the meeting is much more hit and miss!

Sperm are cells with a mission. They're designed to swim, find an egg and fertilise it. But they have to compete with each other for the job. About 500 million sperm landed in your mother's vagina but each egg only lets one sperm in. The rest die.

### Growing your own body

You began life as a single, fertilised egg cell. This cell split into two, and each new cell divided again. When you arrived in the uterus (womb) a week later your cells had divided eight times and formed a hollow ball.

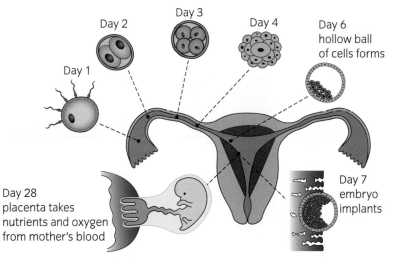

Day 2

Day 3

Day 4

Day 6
hollow ball
of cells forms

Day 1

Day 7
embryo
implants

Day 28
placenta takes
nutrients and oxygen
from mother's blood

A growing body needs food and oxygen. When you settled in the womb your outer cells contacted your mother's blood supply, grew into a **placenta** and began passing nutrients from her blood to yours.

The rest of your cells carried on dividing. They started to specialise and turn into organs – starting with your brain, spine and heart. You began to look human. After 8 weeks growth you were called a **fetus**.

**1**   Work out how many cells an embryo has when it settles in the womb.

**2**   Draw a 1-week-old embryo and label each group of cells.

**3**   Explain how an embryo gets the food and oxygen it needs.

### Learn about

● How sex cells specialise
● How eggs turn into embryos
● How twins form

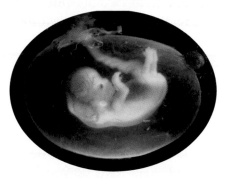

▲ Safe and well fed – an 8-week-old fetus.

## Brainache

**Q** A few women have had four, five, six or seven babies at once. How did that happen?

**A** Fertility drugs made their ovaries release lots of eggs at once. Many of the babies died.

## Multiple embryos

Babies sometimes arrive in twos or threes. An ovary releases extra eggs and they all get fertilised. Most **twins** and **triplets** are no more alike than ordinary brothers or sisters. But **identical twins** form if an embryo splits in two. Each half makes a whole new baby. The split usually happens between days 4 and 9. After 12 days it's too late. A partially split embryo forms **conjoined twins**.

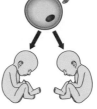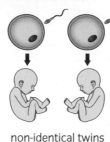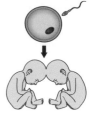

non-identical twins        non-identical twins        conjoined twins

**4**    Explain why some twins are identical and others are not.

## The inside story

Sperm and eggs are both well designed for the job they do.

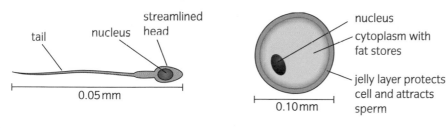

**5**    What features make sperm cells good at swimming? Why do they need these?

**6**    What job does the layer of jelly around an egg do?

**7**    Cilia in the oviducts push eggs towards the womb. Why are they needed?

When a sperm enters an egg it leaves its tail behind and the egg membrane seals up behind it. The sperm nucleus fuses with the egg nucleus to make one – which is copied each time the cell divides.

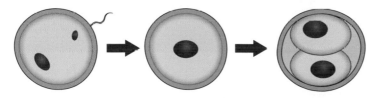

## Summing up

**8**    If a cell just kept dividing, it would make a big round blob. What else happens as the embryo turns into a fetus?

**9**    If the embryo fails to implant it soon dies. Suggest why.

**10**   Yusra had triplets – two identical girls and a boy. How could that happen?

**11**   Twins are more likely to be born prematurely and need expensive medical care. Suggest why.

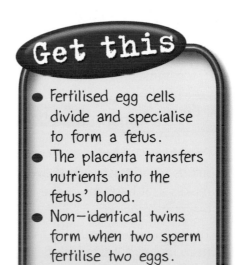

### Get this

- Fertilised egg cells divide and specialise to form a fetus.
- The placenta transfers nutrients into the fetus' blood.
- Non-identical twins form when two sperm fertilise two eggs.
- Identical twins form when embryos split into two.

### Learn about
- Development in the womb
- Things that harm a fetus

### A mind of his own

As Jack grew in his mum's womb, his mind began to form too. His brain, spine and nerves were some of the first parts to grow. When his muscles formed his brain and nerves made them work. He practised all the movements he would make when he was born.

Jack spent 38 weeks in the womb. By his 20th week, he had a full set of **sense organs** – eyes, ears, touch, smell and taste. These were starting to send signals to his brain about the world around him.

At first his brain couldn't make sense of all the signals. Scientists think brains turn on gradually – like lights controlled by dimmer switches. It's a two way process. Signals from Jack's sense organs helped his brain cells to connect.

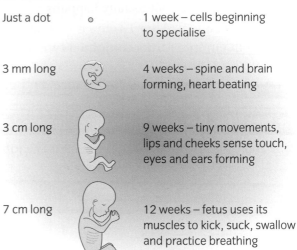

| | | |
|---|---|---|
| Just a dot | o | 1 week – cells beginning to specialise |
| 3 mm long | | 4 weeks – spine and brain forming, heart beating |
| 3 cm long | | 9 weeks – tiny movements, lips and cheeks sense touch, eyes and ears forming |
| 7 cm long | | 12 weeks – fetus uses its muscles to kick, suck, swallow and practice breathing |

**1** We say a normal pregnancy lasts 40 weeks but time this from 2 weeks before the egg is fertilised. How long does a baby actually spend in the womb?

**2** How many weeks pregnant is the mum when the embryo is a week old?

**3** When does a fetus start to move?

**4** Which sense is the first to be used?

**5** How long is it before a fetus responds to sounds?

**6** Why do you think the brain and nerves form first?

▲ 20 weeks – the fetus responds to touch, smells, tastes and sounds, makes walking movements and blinks.

### How do we know... what sounds a fetus hears?

Psychologists study behaviour to find out more about our minds. 4D ultrasounds show them every movement a fetus makes as it happens. If the fetus hears a new sound, it usually reacts by moving.

They can also measure its heartbeat. When it hears something new, its heart beats faster.

**7** A 36 week-old fetus recognises her mother's voice. How might it respond if someone new started talking?

## A good start in life

Jack's mum ate well when she was pregnant, and didn't smoke. Smoking puts an addictive drug called **nicotine** in your blood and cuts the amount of oxygen it can carry. Babies that don't get enough oxygen are more likely to be born early, before 38 weeks, and are called **premature** babies.

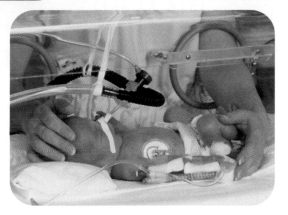

Premature babies find it difficult to breathe and stay warm. They need to be kept in an incubator until they grow stronger.

**8** Which chemical gets into your blood if you smoke?

**9** What is different about babies whose mothers smoke?

High alcohol levels in the blood can damage embryos. They usually survive, but may have learning difficulties or mental health problems.

Doctors used to say a little alcohol was harmless and relaxing – that's 1 or 2 units per week. But the invention of 4D scanners changed that. They allowed psychologists to spot behaviour changes in the fetus when its mother drank. Now pregnant women are advised not to drink at all.

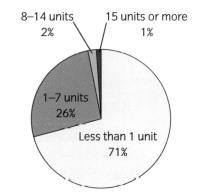

The pie chart shows how much pregnant women drink per week:

**10** How much alcohol do most pregnant women drink?

**11** What percentage drink more than 7 units per week?

## Effects of smoking

The bar chart shows how smoking affects the chance of having a small baby:

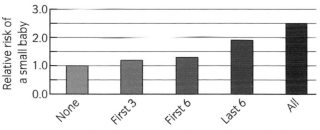

Months of pregnancy when mother smoked

**12** How are birth weights affected when mothers smoke?

**13** When does smoking have the biggest effect?

**14** Chevaz is 3 months pregnant. She has decided to give up smoking. Has she left it too late?

### Summing up

**15** List the changes in the embryo between 1 and 9 weeks.

**16** What changes take place between weeks 9 and 12?

**17** Suggest why pregnant women can't take some medicines.

**18** Women with irregular periods can take 6–8 weeks to realise they are pregnant. Why might that be a problem?

## Get this

- The mind of a fetus develops as its body grows.
- A fetus looks fully formed by 20 weeks.
- A fetus is damaged by nicotine and alcohol.

## Born sociable

Newborns recognise faces and voices. They try to copy the expressions on the faces they see. Jack spent the first hour of his life quietly staring at his mother's face. It helped to establish a bond between them.

Babies communicate with their carers long before they can talk. They can cry for attention, or charm those around them with gurgles and smiles.

**1** Which sense is used for the first time after a baby's birth?

**2** At first a new baby can only see things 25 cm away. What do newborns prefer to look at?

**3** How do these behaviours help a baby to survive?

**Learn about**
- How offspring survive
- How psychologists work
- External fertilisation

▲ Babies learn by imitating what they see other people doing.

### How do we know... what a newborn can do?

Mothers let psychologists study their newborn babies. One made faces at the babies and videoed them. The others decided which expression the babies were copying.

They also gave them dummies connected to computers. Changing their sucking speed made different sounds or images appear. Babies only took 10 minutes to learn how to 'pick' their favourite ones.

**4** How did babies 'pick' things?

**5** Why did the psychologists video the babies' reactions when they made faces at them?

**6** Do you think they should have videoed the faces the psychologist made as well?

**7** Which experiment do you think would give the most valid results?

## Ready to run

Some animals can't lie around sleeping all day – they'd get eaten. They have to be ready to run when they are born.

Zebra foals weigh 65 kg. The first thing they do is stand up, move towards their mother's smell, and follow her if she starts to move. They can run within an hour of birth.

Lions prey on zebras. Lion cubs weigh about 1 kg when they are born and are blind and helpless for 3 weeks. Their mothers keep them hidden for 2 months until they can run. Then they spend 8 months feeding and training them until they can hunt.

Look at the table of data about prey and predators and answer the questions that follow.

| Animal | Adult mass (kg) | Lifestyle | Days in womb | Number of offspring born at once |
|--------|-----------------|-----------|--------------|----------------------------------|
| Antelope | 40 | prey | 180 | 1 |
| Leopard | 40 | predator | 98 | 2–3 |
| Gazelle | 150 | prey | 280 | 1 |
| Lion | 140 | predator | 100 | 2–5 |
| Zebra | 280 | prey | 375 | 1 |
| Tiger | 290 | predator | 110 | 2–4 |

**8** Which spend longest in the womb – predator or prey animals?

**9** How does this explain the difference between zebra and lion cubs when they are born?

**10** Why is this helpful for prey animals?

**11** Which have more offspring at one time – predators or prey?

**12** What do you think is the reason for this?

**13** Zebras chomp grass all day. Why do lions need more time to learn from their mothers?

## Giving offspring a chance

Most frogs lay thousands of eggs, and fish can make millions. They are fertilised in water. The female lays eggs and the male releases sperm nearby that they mix with the eggs. Sometimes fish don't have one partner – thousands of them just release sperm and eggs in the same area.

By the time the offspring hatch from the fertilised eggs the parents are usually miles away and there is no protection for them. Only a few survive – just enough to make sure that type of animal survives.

**14** Fish only have a lot of offspring when they live in a fish farm. What happens to most of their eggs in the wild?

### Get this

- Human babies are born with behaviours that help them survive.
- We know about this because of the work of psychologists.
- Animals that care for their young produce fewer offspring.
- Externally fertilised offspring are often produced in large numbers.

## Summing up

**15** Why does a baby antelope need to run soon after birth?

**16** Why do frogs produce thousands of offspring at once?

**17** How are human babies similar to those of a zebra or a lion?

**18** Babies remember smells from the womb, so new mums are advised not to shower straight away. How does this make their baby more content?

## Learn about
- Changes during puberty
- Why girls have periods

## A growing boy

Jack is 2 years old and his growth rate is slowing down. That's normal.

But when he hits **puberty** he will suddenly grow a lot taller. His appearance and emotions will change too.

Puberty usually happens between ages 11 and 15. Girls have their growth spurts earlier than boys.

**1** When does puberty usually take place?

**2** Why do girls get taller before boys?

**3** Look below at the other changes puberty causes. What does the process get you ready for?

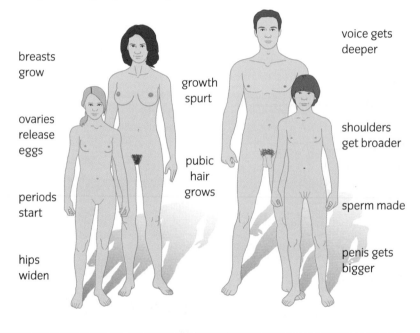

breasts grow

ovaries release eggs

periods start

hips widen

growth spurt

pubic hair grows

voice gets deeper

shoulders get broader

sperm made

penis gets bigger

## Gripped by hormones

When you hit puberty, your brain makes your testes or ovaries pour **hormones** into your blood.

Hormones are chemical messengers which affect every part of your body. You have hormones in your blood all the time, but you only really notice your **sex hormones**. They change your shape during puberty, and they give you a sex drive.

**4** What makes puberty start?

**5** Why do so many things change at once?

## Periods

A girl's periods start during puberty. An egg leaves her ovaries each month and her uterus builds up a thick lining. If the egg is fertilised, the embryo needs a deep layer to implant itself in.

If the egg isn't fertilised the lining breaks down. It leaves her body as a flow of blood through her vagina. Then it all starts again in the next month. The whole process is called the menstrual cycle. It usually lasts 28 days but can be longer or shorter. It can also vary from month to month.

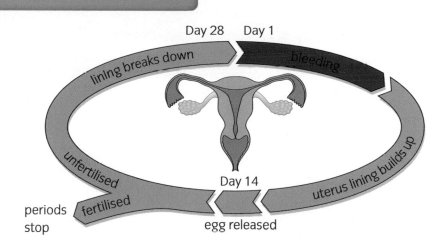

Periods stop while a woman is pregnant and end completely at the **menopause**, when she is about 50 years old.

**6** What makes a girl notice her periods have started?

**7** What happens in the uterus just before an egg is released from one of the ovaries?

**8** If an egg is fertilised by a sperm where does this happen?

**9** What causes the bleeding that takes place during a period?

**10** Why do periods stop when a woman is pregnant?

## Up late?

During puberty your sex hormones affect your mind as well as your body. They can make you feel moody and irritable. They can also upset your body's timer, so it's harder to get enough sleep. It feels like 8 p.m. when it's really 11 p.m. and it feels like 4 o'clock in the morning when it's time to get up for school.

**11** Why do some young people feel tired all the time?

**12** Young people often argue with their parents during puberty. Suggest one thing they might disagree about.

### Summing up

**13** List all the changes sex hormones cause in boys and girls.

**14** In Kyle's Year 9 class there is a wide range of different heights and the girls are mostly taller than the boys. Why?

**15** What are hormones?

### Get this

- Puberty causes:
  - sex organ development
  - rapid height increases
  - the start of periods.
- Sex hormones cause these changes and some side-effects like moodiness.

### A birth in the family

Jack's mum was lucky. Her sister volunteered to help her. But was it as easy as they expected – and would they all do it again?

Having my eggs taken out was the easy bit. They put me to sleep, so I didn't feel anything. Before that I took hormones every day for a month. I had blood tests and scans too. It was all quite stressful. I was fine in the end. But the hormones make some people really ill. I wouldn't want to risk it again.

Jessica

I was worried Jack wouldn't really feel like my son. But as soon as I was pregnant everything changed. My sister donated the egg, but Jack grew in my womb. He's my baby. When he asks where babies come from we'll explain how Jessica helped us.

It did feel a bit weird using Jessica's eggs. She's been great. I wouldn't have wanted to use eggs from a stranger. The clinic has frozen our spare embryos. It cost 2 months wages for the IVF. We were lucky it worked first time.

Jack's mum

Jack's dad

**1**   Jacks parents will tell him about his birth. Some parents who use egg or sperm donors keep it a secret. Why might they think that's the best thing to do?

In the UK, more than 1000 women each year get pregnant using donated sperm. Sperm are easier to collect than eggs. One semen sample contains millions of sperm, but clinics don't let more than 10 women use the same donor's sperm.

## Rules and guidelines

IVF clinics in the UK follow rules and guidelines set by the HFEA (Human Fertilisation and Embryology Authority). But not everyone agrees with their decisions. What do you think?

I agree – we can't encourage women to do something dangerous just for the money.

**RULE**
Women must not be paid for their eggs.

I disagree – women get paid in other countries. They should decide if they want to risk it.

I disagree – egg and sperm donors should be anonymous.

**RULE**
If an adult asks, they must be told where their egg and sperm came from.

I agree – everyone has the right to know where they came from.

I disagree – embryos don't always grow. If they use two there is more chance IVF will be successful.

**GUIDELINE**
Cut the number of twins born by only implanting one embryo.

I agree – twins are often premature. They are expensive to look after and they may have health problems.

I agree – young women make better mothers.

**GUIDELINE**
Women should be under 40.

I disagree – some 50 year olds are still young and healthy. There should be no limits.

**2** With your partner, choose one of these rules or guidelines. Decide who will argue for it, and who will be against it. Find at least **one** more argument for your side of the case. Then take it in turns to put your views to your partner.

**3** Explain why it is important to control new technologies like IVF.

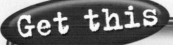

## Get this

New technologies like IVF:
- have advantages and disadvantages for society.
- make things happen which were impossible before
- need new regulations to control them.

# Identifying differences

## Spot the difference

Ben hates broccoli. He says it tastes bitter, and it really does to him. One in four people are like Ben. They can taste bitter chemicals that others can't detect.

We have thousands of features like this which vary.

The differences between us are called **variation**.

**1**    What word describes our differences?

**2**    Why might some people dislike broccoli or olives?

You can recognise hundreds of people – friends, family, people you see only see now and then, and celebrities. It's easy. The shapes of their faces differ and their hair, eye and skin colours vary.

You can spot people you know from a long way off. They have different heights and builds, individual voices and distinct ways of walking.

**3**    The people in the photograph show a lot of variation. List **five** features which vary.

**4**    You notice a friend on the opposite side of a football pitch. How can you tell who it is?

**5**    Your cousin has grown a lot since you last saw her. Her hair is shorter and she has dyed it blond. Why do you still recognise her?

## Hidden variation

Some differences are easy to spot, but most variation is hidden. You might be able to hold your breath for longer, or maybe your heart beats more times per minute. Your blood looks the same as your friend's but it may belong to a different blood group.

Variation makes our bodies work differently. It also affects our behaviour. We prefer different television programs and laugh at different jokes.

There are about 7 billion people in the world. But we all have different combinations of features, so we are all unique.

**6**    Explain why photos can't show all our differences.

## Identification

Ben reads a lot. He changes his library books every week. He used to have a library card. But now he just puts his finger in a scanner. It's very quick and he doesn't have to worry about losing his card.

Everybody's fingerprint has a unique pattern of lines. Even identical twins have slightly different fingerprints.

Using features like fingerprints to identify people is called **biometrics**. Some of our other features are also unique. Most biometric systems use the shape of your face, the patterns on the coloured part of your eye (iris) or your fingerprint. But we also have unique voiceprints and different ways of walking. Since 2007, biometric data has been added to every UK passport.

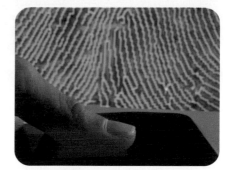

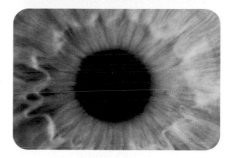

**7** Are identical twins totally identical?

**8** List **three** features that could be used to identify you.

**9** Which feature could identify people over the phone?

**10** The latest fingerprint scanners check for a pulse in the finger. How does this help prevent fraud?

### How do we decide... which biometric feature to use?

Security systems have to be convenient, reliable and as cheap as possible. Iris patterns vary more than fingerprints and face shapes. They change as the iris responds to light, so they can't be faked. But face shapes can be detected from further away. Fingerprint scanners are the cheapest, but they don't work well with dirty or damaged fingers.

**11** Which biometric feature could be used to open your front door as you walked towards the house?

**12** Which would be best to check who needed to pay for a meal from the canteen?

### Summing up

**13** How can common features like hair colour, eye colour and shoe size be used to identify people?

**14** What advantage do biometric security systems have over keys, cards or PIN numbers?

**15** Some people worry that thieves could steal your fingerprints and pretend to be you. Why would this be more serious than having your key stolen?

### Brainache

**Q** Could a bank do without keys, cards or PIN numbers?

**A** In 2007 a Swiss bank became the first to try. It uses face scans and magnetic locks to control entry to the building, and iris scans to limit access to the most secure areas.

### Get this

- Everyone has a different combination of features.
- Biometric systems use unique parts of our appearance or behaviour to identify us.

# What makes us different?

## Inherited variation

Sarah has two little sisters and her mum is about to give birth to a baby. The three girls look different and their personalities are not the same. But they have all **inherited** some of each parent's features.

Sarah's eye colour and blood group won't change as she gets older. But she hasn't stopped growing. She can't be sure how tall she will grow, or what her body mass will be. Some features are present when we are born. Others develop as we grow and age.

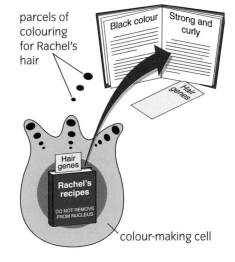

**1** Which of her mother's features does Sarah have?

**2** Why do the sisters resemble their parents?

**3** Name **two** features that are fixed before birth.

**4** Name **two** features that develop as we grow.

## All in the genes

Your features depend on the **genes** you inherit from your parents. You have genes in the nucleus of every cell. They act as your body's recipe book.

Genes tell your cells what to do. Your cells make up the tissues and organs that form your body, so what they do shapes your whole body.

Most cells only do certain jobs. They don't use every gene. Rachel has some cells which just produce the colour for her hair. They follow the instructions from her genes for hair colour.

Some genes are the same in everyone – so we all have the same basic body plan. But we inherit different versions of other genes.

parcels of colouring for Rachel's hair

Black colour | Strong and curly

Hair genes

Hair genes

Rachel's recipes

DO NOT REMOVE FROM NUCLEUS

colour-making cell

**5** Explain how Rachel's genes made her hair black.

**6** Why don't your cells use every gene?

## Mixing it up

Sperm and egg cells carry a random selection of genes from the person making them. When a sperm fertilises an egg the baby that forms has a unique mix of genes.

So Sarah and her sisters inherited different combinations of their parents' genes. This makes them similar but not identical. They show **inherited variation**. Identical twins are the only people with exactly the same genes because they grow from a single egg and sperm.

Scientists think we have around 25 000 genes. That can make a lot of unique combinations!

**7** An egg has been fertilised. It starts to divide and grow. What decides the features the new baby will have?

**8** Rachel and her sisters are similar, but not identical. Explain why.

**9** Sarah's dad thinks the new baby will inherit his skill at football, and his wife's musical ability. Explain why he might be wrong.

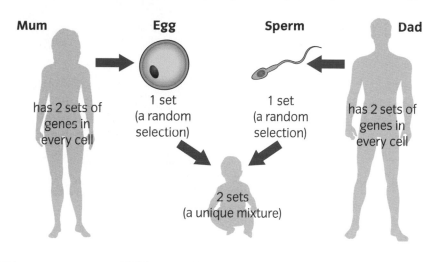

**Mum** has 2 sets of genes in every cell

**Egg** 1 set (a random selection)

**Sperm** 1 set (a random selection)

**Dad** has 2 sets of genes in every cell

2 sets (a unique mixture)

## Limited options

Everyone inherits a unique combination of features, but each feature may have a limited range of options.

We all have one of the four main blood groups: O, A, B or AB. Differences like this show **discontinuous variation**. There are only a few different options and we all have one of them.

Hair colours are decided by more than one gene, so they show a lot of variation. But they can be sorted into a few main groups. The percentage in each group varies. Black is the world's most common hair colour, but most people in the UK have brown hair.

**10** Which blood group is most common in the UK?

**11** The bar chart shows the hair colours in a class of students. What percentage had black hair?

**12** Would you expect the same result in another country?

## Summing up

**13** Why don't people look exactly like their mother or father?

**14** Can you predict the features your children will inherit?

**15** Which of these show discontinuous variation?

height   blood group   shoe size   finger length

Blood groups in the UK

Hair colours in the UK
Red   Brown   Blonde   Black

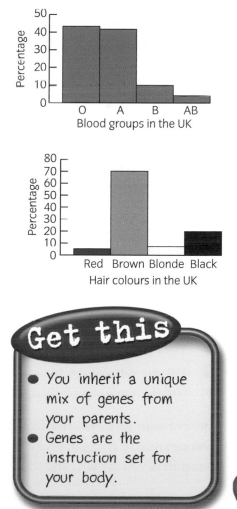

**Get this**

- You inherit a unique mix of genes from your parents.
- Genes are the instruction set for your body.

35

## Size control

Sam's family don't have a garden but his mum grows miniature trees called bonsai. This bonsai has he same genes as a full-sized oak tree but the shallow tray stops its roots growing. Oak trees normally grow 100 times bigger.

Some differences are not caused by genes. Plants that don't get enough light, water and minerals don't grow as big as those that do. This is **environmental variation**.

**1** List **two** things that can affect a tree's growth.

**2** If a seed from the bonsai was planted in a large garden, would it grow into another miniature tree?

**3** The lower leaves on a tree are often thinner and paler than those near the top. Which difference in their environments could cause this?

## Environmental variation

Humans and other animals are affected by their environments too. Your environment doesn't just mean your surroundings. Everything you eat, drink, do and learn affects you in some way.

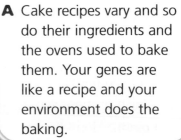

Sam has an identical twin brother called Tony. They inherited the same genes from their parents.

They were hard to tell apart when they were babies. But they were in different classes at school and their parents always bought them different clothes and toys. Tom was ill last year and lost a lot of weight, but he still plays football. Sam prefers swimming.

**4** What features have the twins inherited that are the same?

**5** Why are identical twins more alike when they are babies?

**6** List the differences between the boys now.

**7** Where do you think these differences have come from?

Most of your features are affected by genes and your environment. You can inherit the genes for a fit, athletic body; but you won't get one unless you eat well and train hard.

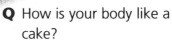

**Brainache**

**Q** How is your body like a cake?

**A** Cake recipes vary and so do their ingredients and the ovens used to bake them. Your genes are like a recipe and your environment does the baking.

## How can we tell... that our environments affect us?

Identical twins have the same genes. Ordinary twins don't. If identical twins living apart show close similarities in a feature, it's probably inherited. If they show big differences, it's probably caused by differences in their environments. Comparing ordinary and identical twins brought up together provides extra evidence.

| Twins measured | Height difference in cm | Mass difference in kg |
|---|---|---|
| 50 identical pairs | 1.7 | 1.9 |
| 50 ordinary pairs | 4.4 | 4.5 |
| 20 identical pairs living apart | 1.8 | 4.5 |

The table shows the **average** differences between pairs of twins. In each group, all the results were added together and divided by the number of pairs.

**8** From this data, does height variation seem to be inherited or environmental?

**9** Does the variation in mass seem to be inherited, environmental or a bit of both?

**10** How could you compare ordinary and identical twins fairly?

## A bit of both

Environmental factors can affect inherited features. Your height is usually an inherited feature. But you only reach the height programmed by your genes if you eat nutritious food and get enough rest and exercise.

In 2001 the average height of 30 year-old women was 165 cm. That's 6 cm taller than the average woman's height in 1951. This isn't because their genes have changed. We have a better standard of living now and more nutritious food. More people are reaching the height they should be.

**11** People whose diets lack minerals are often shorter than average. Is this an example of genetic or environmental variation?

**12** In 2001 the average woman in the UK had a mass of 65 kg. In the USA the average was 71 kg. Suggest **two** things that could account for the difference.

1951        2001

## Summing up

**13** What are the two main sources of variation?

**14** Decide whether each of the following is caused by genes, the environment or both: language/s spoken   fingerprint pattern   blood group   skill at football   eye colour   body mass   health.

**15** Some people's ears stick out. They can have surgery to pin them back. Does this stop their children having the same problem?

**Get this**

- Nearly everything about you is influenced by your environment, or a combination of your genes and environment.

Shalini loves shopping. But it takes her hours to find clothes that fit. None of the labels has her exact measurements on it.

Few people have exactly the size on the label. Body measurements, like height, show **continuous variation**. They can take any value within a certain **range**.

Features like height are influenced by many different genes and a variety of environmental factors. Those that show discontinuous variation, like the main blood groups, are only influenced by one or two genes.

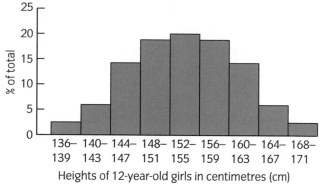

### Learn about
- Continuous variation
- Correlations

**1**  Foot lengths are influenced by many different genes. Are they likely to show continuous or discontinuous variation?

**2**  What sort of variation do shoe sizes show?

The average height of a 12-year-old girl is 153.5 cm in the UK. You find it by adding the heights and dividing by the number of girls. To get an accurate value you need to measure a large number of girls from all over the country.

**3**  Shalini is 158 cm tall. Her friends' heights are 150 cm and 151 cm. Is each girl taller or shorter than the UK average for 12-year-old girls?

**4**  Work out their average height.

**5**  According to the graph, what range of heights is most common?

**6**  What percentage have heights between 168 and 171 cm?

Heights of 12-year-old girls in centimetres (cm)

### How do we know... what sizes people are?

Between 2001 and 2004, SizeUK used 3D whole-body scanners to measure 11 000 people from different adult age-groups. Each scan took a few seconds. It let a computer take 130 body measurements at once. Shops are using the data to make sure their clothes match the sizes of their customers.

**7**  Why did they check so many people?

**8**  People usually stop getting taller before they are 20. Suggest another body measurement that might change with age.

## Getting the right size

The sleeves on Shalini's new jacket fit perfectly. Jackets only have one measurement on the label – the wearer's height, age or chest size. Clothes makers try to make the arm lengths right for each size.

The top **scatter graph** shows the heights and arm lengths of some 12-year-old girls. Most of them lie close to a diagonal line. This means there is a **correlation** between their heights and their arm lengths. If the correlation was perfect all the points would be on the line.

**9** The two shortest girls are 143 cm tall. How long are their arms?

**10** Do taller girls have longer arms?

**11** Three of the girls were 151 cm tall. Are their arms the same length?

**12** Abina is 160 cm tall. How long would you expect her arms to be?

**13** Maria is 150 cm tall and her arms are 50 cm long. What problem might she have with jackets?

**14** Suggest another body measurement that might show a correlation with height.

The second scatter graph shows the same girls' head circumferences. The points do not lie close to an obvious diagonal line. So there is **no correlation** between the two variables.

**15** What were their largest and smallest head circumferences?

**16** The average head circumference was 55 cm. Find the heights of the tallest and shortest girls with this head size.

**17** Could a girl's height be used to predict her hat size?

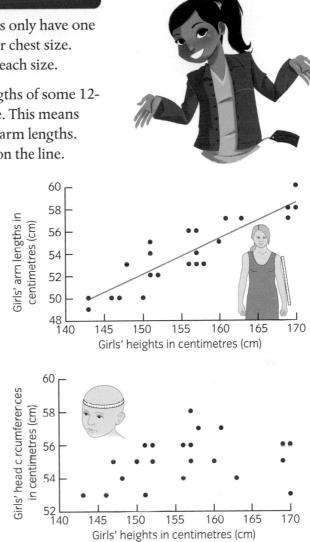

### Summing up

**18** What is continuous variation?

**19** Which of these measurements show continuous variation?

arm length   shoe size   body mass

**20** Explain why clothes makers can predict some of your other body measurements if they know your height.

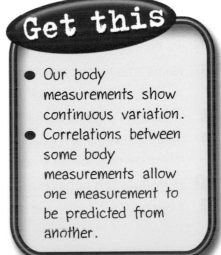

### Get this

- Our body measurements show continuous variation.
- Correlations between some body measurements allow one measurement to be predicted from another.

39

### Born clever?

Animals often seem to know what to do from the moment they are born. This bird is a cuckoo. Its mother laid her egg in another bird's nest. As soon as it hatched, it heaved all the other eggs and chicks out of the nest and started calling for food. The cuckoo's foster parents are tiny compared with a cuckoo. They will struggle to find enough insects to feed it.

**1** What does a cuckoo chick do as soon as it hatches?

**2** Why does it have to kill to survive?

**3** Why don't its foster parents leave it to die?

Every animal inherits some **instinctive behaviour patterns**. They are triggered by **signals** they sense in their environment.

### How do we know... signals trigger instinctive behaviour?

**Ethologists** study the way animals respond to their environments. They watched gull chicks peck at their parents' beaks soon after they hatched. The pecking made their parents feed them. But the chicks could be fooled by simple model beaks.

**4** Which 'beak' were they least interested in?

**5** Which feature acted as a signal to peck?

| Model | | Relative number of pecks |
|---|---|---|
| realistic beak | | 100 |
| beak with no red marks | | 35 |
| red pencil | | 126 |

### Having a strategy

Woodlice live under logs and stones. They dry out in the sun and die. These woodlice have just been disturbed. They are moving all over the place, but they follow a few simple rules.

**6** Name **one** thing the woodlice can detect.

**7** How do they move over dry ground?

**8** How does their movement differ over damp ground?

**9** Imagine woodlice moving across a piece of filter paper. Will they take longer to cross damp or dry paper?

**10** How does this behaviour help the woodlice?

**11** How might they behave if they sensed light ahead?

| Movement | Ground | |
|---|---|---|
| | damp | dry |
| speed | slow | fast |
| turns | often | rarely |

## Learning fast

Behaviour patterns that help animals survive their first few days are largely inherited. But animals are born ready to learn some things quickly – like who their parents are and what food they should eat. This behaviour is called **imprinting**. It is difficult to reverse and only occurs during a narrow period of time. This boy was the first large moving object the ducklings saw. So they assumed he was their mum, and they won't stop following him around.

 **12** Why are the ducklings following the boy?

**13** Is their behaviour caused by their genes, environment, or both?

**14** Why is this sort of behaviour useful?

## Linking experiences

Most behaviour is less predictable. It is influenced more by experience than genes. Animals can learn to link behaviour they were born with to new signals. Pavlov, a Russian biologist first noticed this with dogs.

Dog smells food and makes saliva.

Dog hears bell and ignores it.

Dog hears bell and smells food and makes saliva.

Dog hears bell and makes saliva.

Sit!

 **15** What does a dog do when he smells food?

**16** How does the dog learn to make saliva when he hears a bell?

## Working for rewards

Dog trainers know that animals learn quickly if they get a reward, or avoid something bad. The training works best if the dog gets the reward around 1 second after he shows the right behaviour.

 **17** Your dog sits when you shout 'SIT!'. What should you do to make him repeat this behaviour?

### Summing up

**18** Say whether each of the following is caused by an animal's genes, their environment or both: cuckoos push other eggs out of the nest; gulls peck a red pencil; woodlice run straight across dry paving stones; orphaned ducklings follow a dog.

**19** Explain why a dog might run into the kitchen when they hear you opening a can of beans.

## Get this

- Inherited behaviour patterns help animals to survive.
- But most behaviour depends on the way their genetic and environmental influences interact.

**How Science Works**

### The fastest thing on no legs

Oscar Pistorius has no feet but has won gold medals in 100 m, 200 m and 400 m races against other disabled athletes. He runs on carbon fibre blades. Some say he has an unfair advantage over 1-legged runners, because he could vary his height. Others say he shouldn't race 2-legged runners because his blades might give him an advantage.

The graph shows the average heights of the fastest runners at each distance between 1990 and 2003.

### Learn about
- How variation affects athletic ability

| Year | Olympic 100 m gold medal winners | Height (cm) |
|------|----------------------------------|-------------|
| 1988 | Carl Lewis, USA | 188 |
| 1992 | Linford Christie, GBR | 191 |
| 1996 | Donovan Bailey, CAN | 182 |
| 2000 | Maurice Greene, USA | 176 |
| 2004 | Justin Gatlin, USA | 185 |

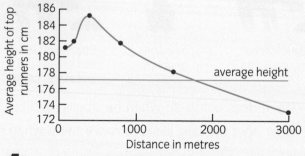

**1** Over which distance does being tall seem to give you most advantage?

**2** The heights on the graph don't include shoes. Estimate how much extra they would add.

**3** What distances does Oscar run?

**4** Oscar stands 187 cm tall in his blades. Might that give him an advantage?

**5** Most people aren't average. Have taller men than Oscar won the 100 metres?

**6** Can shorter men compete?

### Hidden advantages

Top athletes have to move fast. They need efficient muscles. We have two main types of muscle fibre. Slow-twitch fibres can keep going for longer. Fast-twitch fibres are more powerful, but tire easily. When scientists tested different groups of athletes they found different combinations of the two.

**7** What combination of muscle fibres does the average person have?

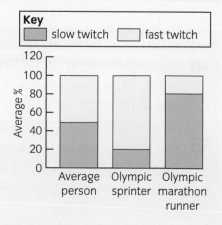

**8** What is different about marathon runners?

**9** How does the combination found in sprinters suit them to their sport?

## Working harder

Slow-twitch muscles need a good oxygen supply. In the 1960's, Finnish skier Eero Mäntyranta discovered he had a sporting advantage.

**10** What does the table tell you about Eero's blood?

**11** Why did this mean he had a better oxygen supply?

|  | Red cell count in billions per litre |
|---|---|
| normal range | 4.5–5.6 |
| Eero's blood | 6.7 |

Years later scientists were able to show that he'd inherited a rare gene. A drug used to treat anaemia can cause a similar effect. Athletes are banned from taking it. But they are allowed to train at high altitudes, or by breathing air with less oxygen in it, and these both increase red cell counts.

**12** Why could Eero's muscles work harder?

**13** What made his blood so different?

**14** How can athletes legally increase their red cell counts?

▲ A controlled air supply simulates conditions 4000 m above sea level for these footballers.

## Being the best

Scientists have discovered more than 100 genes that affect athletic ability. But psychologist Karl Ericson says genes aren't everything. Everyone with talent he studied had been doing deliberate practice for around 10 years with good coaches. Oscar Pistorius started to win races straight away. But he had been practicing other sports for years.

**15** According to Ericsson, what's the best way to achieve success?

**16** Why is it important to enjoy your sport?

Ericson's advice

1 set goals

2 practice

3 listen to feedback

4 improve weaknesses

Repeat for 10 years.

## Talent spotting

In 2007, the UK sports agency began to fast-track extra athletes to qualify for the 2012 Olympics. They looked for athletes who might do better in new sports, and younger athletes who could benefit from training. They set themselves the ultimate goals of gaining fourth place in the 2012 Olympic medal table, and first in the Paralympic medal table.

**17** Why did they look for athletes who were already good at other sports?

**18** What features might a good 400 m runner have?

**19** What would make you think someone could run a marathon?

**Get this**

- Inherited features and training both influence athletic ability.

## Learn about
- Species
- What affects their numbers

## Too late!

In December 2006, 30 scientists spent six weeks searching China's Yangtze River for dolphins like this. But they didn't find one. There may be none left to find. The Yangtze dolphin could be **extinct**.

Most dolphins live in the sea but this was a river dolphin. It could not survive in salty sea water. It was a unique type of dolphin found nowhere else on Earth.

Each type of animal or plant is a **species**. Nearly 2 million species have been found and biologists estimate there are at least 5 million. This vast range of species is called **biodiversity**. But many living things are struggling to survive. Many could go extinct before they are even named.

**1** What does it mean if an animal is declared extinct?

**2** The Yangtze river was very polluted and there were few fish left for dolphins to feed on. Why couldn't they move to a better home?

## Species

Members of a species can look very different. Like humans, they show a lot of variation.

These dogs are all one species. Any male dog can mate with any female dog. Their offspring will grow up and have puppies of their own. If organisms can have offspring they are **fertile**.

Members of the same species only produce fertile offspring if they breed with each other – not with other species.

**3** Do members of a species always look the same?

**4** How could you prove two groups of animals belonged to the same species?

This female will never have offspring. She is a mule. Her mother is a horse and her father is a donkey. Different species can sometimes mate, but their offspring usually can't. Most are **infertile**.

**5** How are mules produced?

**6** Mules are stronger and more intelligent than horses. Why aren't they a separate species?

## Changing populations

This species is not likely to die out. There are more than 60 million of them in the UK. Rat numbers are rising and they are a serious **pest**. They spread disease, ruin crops and food stores, and chew through cables.

**7** Are rats in danger of becoming extinct?

**8** Why are rats a serious pest?

**Herbivores** eat plants and **carnivore**s eat animals, but rats are omnivores – they eat anything. So rats are part of many different food chains. The arrows in each chain point from the food to the animal that eats it. The animals that eat rats are their **predators** and rats are their **prey**.

**9** Choose a prey to complete this food chain:
? → rat → eagle

**10** Choose a carnivore to complete this chain:
waste food → rat → ?

**11** Name **three** other predators that attack rats.

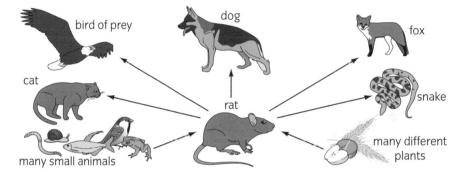

A warm, well-fed rat can give birth once a month, to six young at a time – and her daughters start mating when they are 3 months old. So rat numbers go up fast where there are no predators.

**12** Takeaway food is often dropped in the street or put in open bins. How might this affect rat numbers?

**13** How are rats affected when stray cats and dogs are removed from the streets?

**14** Rat poison harms birds of prey. Why do we have to use more poison now than we used to?

### Get this

- Each type of animal or plant is a species.
- Members of a species can produce fertile offspring.
- A species' numbers increase if they have plenty of food and few predators.

### Summing up

**15** What is a species?

**16** Why can't we say exactly how many species there are?

**17** List **two** things that affect animal numbers.

**18** Most rat predators will also eat squirrels. How might the numbers of squirrels be affected if all rats were poisoned?

**19** Rats often hide in boats, but their predators don't. The rats can swim ashore when the boat is near land. What problems can this cause?

## Sorting things out

Olivia wants some satsumas, so she heads for the fruit section and looks for the oranges. There are several types of oranges there, including satsumas.

Supermarkets put similar things together to make them easier to find. This is **classification**.

Scientists classify species in a similar way. Putting them into groups makes it easier to store what we know about them and give each species a different name. Scientists everywhere use the same system so they can share data.

**1** What does classification involve?

**2** Why is it important for all scientists to name and group species in the same way?

**Learn about**
- Classification
- Invertebrates

## Grouping living things

Most species can be classified as plants or animals. The difference between them is that animals feed on other living things while plants make their own food using photosynthesis.

**3** What is the same about all plants?

**4** Use the pie chart to find the percentage of species classified as plants or animals.

Plants and animals are each divided into smaller subgroups.

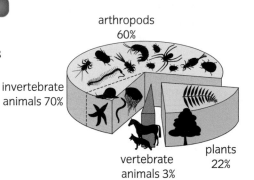

arthropods 60%

invertebrate animals 70%

vertebrate animals 3%

plants 22%

## Animal subgroups

For animals, the first division is between **vertebrates** and **invertebrates**. Vertebrates have bony skeletons with **backbones** inside their bodies. Invertebrates don't.

Sometimes it's hard to tell which animals have backbones from the outside. A roundworm and snake look similar. Roundworms are invertebrates, but the X-ray on the right shows the snake has a backbone – it's a vertebrate.

**5** Which group has most species – vertebrates or invertebrates?

**6** How are vertebrates different from invertebrates?

## Groups within groups

Vertebrates and invertebrates can both be sorted into smaller subgroups based on what they look like. Most invertebrates belong to one subgroup – the **arthropods**.

The name arthropod means 'animals with jointed legs', but they also have hard outer skeletons. All the animals in this photograph are arthropods.

**7** How can you tell an animal is an arthropod?

**8** Use the pie chart opposite to find the percentage of species classified as arthropods.

There are more than a million arthropod species. Many have the same number of legs. So this can be used to split them into smaller groups. Most arthropods are insects – animals with six legs.

Nearly a million insect species have been named. A third of these insects have tough covers over their wings – they are **beetles**. Five thousand of the beetle species are called **ladybirds**. They have brightly coloured, rounded bodies.

Many ladybirds are red with black spots, but only one species has exactly seven spots.

Having all these subgroups makes it easier to devise **keys** to identify each species.

**9** How can you tell an arthropod is a beetle?

**10** List all the groups the 7-spotted ladybird belongs to. Start with the smallest.

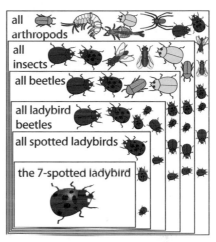

arthropod — does it have 6 legs? yes → insect / no → other arthropods

insect — does it have a tough cover? yes → beetle / no → other insects

beetle — does it have a brightly coloured, round body? yes → ladybird / no → other beetles

ladybird — is it red with spots? yes → spotted ladybird / no → other ladybirds

spotted ladybird — does it have 7 spots? yes → 7-spotted ladybird / no → other spotted ladybirds

all arthropods / all insects / all beetles / all ladybird beetles / all spotted ladybirds / the 7-spotted ladybird

### Summing up

**11** What are the **two** biggest groups of living things?

**12** What are the **two** main groups of animals?

**13** Why is it useful to divide insects into smaller groups?

**14** A biologist once said, 'the Creator must have an inordinate fondness for beetles'. What might make him think that?

**Get this**

- We classify species into groups and subgroups to help identify them.

**Learn about**
- Vertebrates

## Animals with backbones

Vertebrates are bigger and smarter than most invertebrates. They have a complicated nervous system. A skull keeps their brain safe and a backbone protects their spinal cord. This one is more than 20 metres long. It belonged to a whale.

There are fewer vertebrates than invertebrates, but it is still helpful to classify them. Vertebrate species form five main groups. The features each group shares are listed in the chart below.

**1** What job does a vertebrate's backbone do?

**2** How many groups of vertebrates are there?

## Vertebrate groups

**Fish** are only found in water. Their gills don't work in air. **Amphibians** can breathe air but only reproduce in water. Other groups have species that can live in drier places. They have waterproof body coverings and use internal fertilisation.

**vertebrates**

| fish | amphibians | reptiles | birds | mammals |
|------|-----------|----------|-------|---------|
| • gills | • eggs laid in water | • waterproof eggs | • shelled eggs | • give birth |
| • scales | • larvae have gills | • hard scales | • feathers | • produce milk |
| • fins | • naked skin | | • wings | • have hair |

**Mammals** and **birds** eat more for their size than species in the other three groups. The extra energy keeps their body temperature around 37 °C – so they are warm-blooded. Fish, amphibians and **reptiles** are cold-blooded. Their temperatures stay the same as their surroundings. But reptiles can warm up by basking in the Sun.

**3** What makes amphibians and fish different?

**4** A pet snake only needs feeding once a week. Most cats and dogs eat twice a day. Why do they need more food?

**5** Bats are flying mammals. They sleep during the day, so they are rarely seen. List the features you would expect them to have.

## Amphibians and reptiles

Amphibians and reptiles can be hard to distinguish from a distance. Both these animals have four legs and a tail. The salamander is an amphibian. Its skin is smooth and moist and it lives in cool, damp places. Its young develop in water so they have gills instead of lungs.

The gecko is a reptile. Tough scales protect its skin and it lays its eggs on land. Some reptiles live in or near water, but geckos like hotter, drier environments.

**6** Amphibians and reptiles were originally put in the same vertebrate group. What is similar about them?

**7** Now list the differences between them.

**8** Caecilians live in damp soil in the tropics. They look like giant worms but they have backbones. Their skin is moist and slimy. Which group would you put them in?

## Unusual mammals

Whales are unusual mammals. They have very little hair and their bodies are streamlined. But they give birth to live young like other mammals. The newborn calves are pushed to the surface to take their first breath. They feed on their mother's milk for at least 6 months.

**9** What is the same about whales and fish?

**10** Now list some differences between them.

**11** What features do whales and humans share?

## How do we tell others... what we know about species?

There are thousands of books, museum displays and web pages about plants and animals. In 2007, biologists and computer experts began to link this data into a multimedia 'Encyclopedia of Life' accessed from a single web page.

**12** Why is it difficult to find information about some species?

**13** What are computer experts doing to help?

**14** Give **one** reason why biologists want to know about every species on Earth.

## Summing up

**15** Say what's special about each group of vertebrates.

**16** Young children classify animals as 'living on land' or 'living in water'. Name **two** very different vertebrates in each group.

**17** Porpoises are mammals. They live in the shallow seas around the coast. List the features you would expect them to have.

## Get this

- Vertebrates are grouped according to:
  - how they reproduce
  - their body covering.

49

**Learn about**
- Plant groups
- Naming species

### Not just for eating

We eat plants and feed them to farm animals. We turn them into useful products like paper, build homes and furniture, and burn them as fuels. But some plants have specialist uses. Fifty percent of our medicines originally came from plants.

Natives of the Southeast Asian forests use 6500 different plants as medicines and few are known elsewhere. Special plants like these need to be identified and protected.

**1** List **five** things we use plants for.

**2** Give **one** reason why rare tropical plants should be protected.

### Flowering plants

Most useful products come from flowering plants. Flowers come in a huge variety of shapes, but related plants have similar types of flower. So these can be used to divide them into subgroups.

Grasses and trees often have small, dull flowers. Their pollen is spread by the wind so they don't attract insects.

**3** Why are grass flowers sometimes hard to spot?

**4** What makes flowers useful for classification?

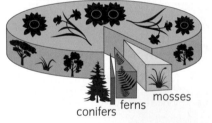

flowering plants

conifers  ferns  mosses

### Plants without flowers

The three plants in the photo don't make flowers. The **moss** on the ground has no roots or veins. The **ferns** under the trees grow taller. They have veins to carry water from their roots to their leaves.

Mosses and ferns only reproduce when they are wet – and they make **spores** instead of **seeds**. The trees are **conifers**. Conifers and **flowering plants** both make seeds but conifers have cones instead of flowers. Seeds can be produced in dry conditions and may survive for thousands of years if kept cold and dry.

**5** Make a key to decide which group a plant is in. Start with:

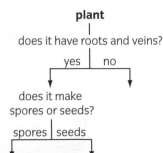

**plant**

does it have roots and veins?

yes | no

does it make
spores or seeds?

spores | seeds

## oceans

Most of the plants in the sea are tiny **phytoplankton** – plants with one cell or just a few. These have been magnified about 500 times. There are thousands of different species. They make their own food, using carbon dioxide dissolved in seawater, and release oxygen. Animals feed on them, so they are vital to life in the sea.

**6** What are phytoplankton?

**7** Why are they important?

## How do we know... how to name species?

In 1735, a Swedish plant expert called Carl Linnaeus invented the system we still use. He sorted plants into groups and subgroups and gave each species two names.

The first name shows their smallest subgroup, and the second shows their species. The plant to the right is *Aloe vera* which means the true aloe. The subgroup aloe contains all the aloe species, but this is the one with real medicinal properties. The names are in Latin which was the international language of science at the time.

Linnaeus was good at getting publicity. His students enjoyed his lessons, he organised popular plant-finding trips and he wrote lots of books. So everyone started to use his system.

**8** Why is Latin used for scientific names?

**9** List **three** things Carl Linnaeus did that got publicity for his system.

**10** Which part of a scientific name is the same for similar species?

## Summing up

**11** Plants are important. Name some things we get from them.

**12** List the four main groups of land plants.

**13** A newly discovered plant is 1 m tall, has feathery leaves and never flowers. Which group would you put it in?

**14** Surtsey is a new volcanic island near Iceland. Soon after it formed, plants began to grow on the bare rock. The plants are only a few mm tall and have no roots. What are they?

**15** What is the most common type of plant in the oceans?

## Get this

- Land plants include:
  - flowering plants
  - conifers
  - ferns
  - mosses.
- Most of the plants in the oceans are microscopic.

### Skeletons from the past

We all belong to the same species – *Homo sapiens*. So people from any country can have children together, and their children will be fertile.

But 30 – 35 000 years ago there were two species of human in Europe. We know this because two sorts of fossilised human bones have been uncovered from this period.

**Learn about**
- How evidence can support scientific ideas or prove them wrong

The skull on the right is from one species. It looks just like a modern skull – and so do the rest of its bones. The other has shorter, stronger bones and a bigger skull – the one on the left. This was *Homo neanderthalensis* or 'Neanderthal Man'. The species seemed to disappear 30 000 years ago – 5000 years after *Homo sapiens* moved into Europe.

**1** Why do we think that human bones buried 35 000 years ago come from two different species of human?

| Species | Homo sapiens | Homo neanderthalensis |
|---|---|---|
| Skull size in cm$^3$ | up to 1500 | up to 1700 |
| Average height in cm | 175 | 168 |

**2** Humans vary a lot. Give **two** pieces of evidence that show modern humans are all one species.

### New bones

In 2003, bones were found on an island near Australia called Flores. The mud they were buried in was 18 000 years old. They fitted a skeleton 100 cm tall – the height of a modern 6-year-old. Her skull measured only 400 cm$^3$ – like a newborn baby's – but her worn teeth showed she was 30 when she died.

There were stone tools nearby and the bones of a miniature species of elephant. The bones had cut marks on them and looked burnt.

**3** Why did scientists think they'd found a child's skeleton?

**4** What made them change their minds?

**5** What evidence is there that the bones belonged to someone clever enough to hunt and cook?

There is still a tribe of small-bodied people living on Flores. They are a bit taller than the ancient skeleton and their skulls are much larger.

**6** Look at the diagram. What differences are there between the woman found buried on Flores and the small-bodied women that live on Flores now?

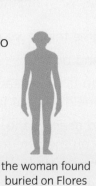

the woman found buried on Flores

an average modern woman

a small-bodied woman from a tribe that lives on Flores now

## The 'new species' theory

The skull from Flores is on the left. It is less than half the size of one from the smallest people living now. So some scientists say it must be from a new species. They call it *Homo floresiensis*.

Tiny versions of other mammals have been found on islands. Small animals survive better where there isn't much to eat.

**7** Why was the skeleton thought to be a new species?

**8** Have other miniature mammal species been found?

**9** Why might small species be common on islands?

## How do we know... what a skeleton's face looked like?

Artists work with scientists to repair the skull and build a face over it. Small bumps on the bones show where each muscle was attached and how strong it was. Then the right depth of tissues is added for the skull's age and sex. But there is quite a lot of guesswork involved.

**10** Can we be sure the woman from Flores looked like this?

## Rival theories

Many scientists argue that *Homo floresiensis* isn't really a new species. Evidence for and against is listed below.

| | | |
|---|---|---|
| For its size, the Flores skeleton had a smaller skull, longer arms and bigger feet than any that have been found before. | The Flores skull is the same size as those from *Homo sapiens* with microcephaly – a rare disease that stops your brain growing. | A tribe of small-bodied people who live on Flores now say they are descended from tiny, hairy folk. |
| Bones from eight other humans who lived on Flores over 12 000 years ago have been found. They were all small. But there is only one skull. | The exact shape of the Flores skull does not match the shapes of skulls from *Homo sapiens* with microcephaly. | The small-bodied people living on Flores now have similar skeletons to the remains. But they are a bit taller and their skulls are bigger. |

**11** Choose evidence that supports the idea that the bones were from a new species of human.

**12** Which evidence suggests the bones were from a diseased member of our species?

**13** What extra evidence would strengthen the 'new species' theory?

## Get this

- Other species of human have existed in the past.

## Amazing acids

Vomit, vinegar and lemons taste sour. Why? They all contain **acids**.

Acids are vital to life. Hydrochloric acid in your stomach helps digest food. Ascorbic acid in fruit – vitamin C – keeps skin healthy, and helps to make bones. You need omega-3 fatty acids from oily fish to defend your body against disease and repair damage.

Many acids are really useful. Orange squash contains an acid to stop it going off. Ethanoic acid (vinegar) preserves pickles.

Acids can also be a nuisance. Methanoic acid makes bee, ant and nettle stings painful. Acids make sweat smelly, too.

**1** Why does vomit taste sour?

**2** Name two acids you must eat. What does your body use these acids for? What foods are they in?

**Learn about**
- Acids and alkalis
- Uses of acids and alkalis

## Alarming acids

Murderer John Haig experimented with dead mice and sulfuric acid. It took only half an hour for the acid to dissolve one mouse. So he left the bodies of his human victims in barrels of sulfuric acid until they turned to sludge. Only the victims' false teeth remained.

▲ This isn't bird poo – it's urine! It's mainly solid uric acid.

Sulfuric acid is a **strong acid**. Strong acids are corrosive – they destroy living tissue. So they can burn your skin and eyes. You might use two other strong acids in science: nitric acid and hydrochloric acid.

Of course, the acids you eat are not dangerous. They are **weak acids**.

**3** The label shows that the acid is corrosive. Describe what will happen if you splash it on your skin.

**4** What safety precautions must you take when using the acid in the picture?

**5** All acids taste sour. But you must never taste a chemical to see if it might be an acid. Why?

HYDROCHLORIC ACID DIL.

CORROSIVE

SURFISIL Safety Coating

No. L11    UK & Overseas Patents Pending

## Using sulfuric acid

Sulfuric acid is hugely important – and not just to murderers!
Worldwide, factories make over 150 million tonnes of it each year.
The pie-chart shows why sulfuric acid is useful.

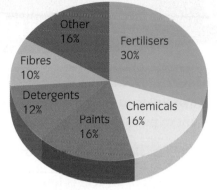

**6** What is the biggest use of sulfuric acid?

**7** Why are fertilisers and detergents useful?

## Astonishing alkalis

**Alkalis** are the chemical opposite of acids. They feel soapy. You
probably use one alkaline substance – toothpaste – every day.
Washing powders and many drain cleaners contain alkalis.

Sodium hydroxide is an important alkali. Companies use it to make
paper, detergents and aluminium for fizzy drink cans.

**Strong alkalis** like sodium hydroxide are extremely corrosive. A small
splash on your skin or eye gives a very nasty blister. **Weak alkalis** –
such as seawater and toothpaste – are not corrosive at all.

BUSTER KITCHEN DRAIN CLEAR CONTAINS
SODIUM NITRATE AND SODIUM HYDROXIDE

CORROSIVE    OXIDISING    HARMFUL    HIGHLY FLAMMABLE

**8** Look at the label. Name the alkali in the drain
cleaner.

**9** Look at the hazard symbols on the drain
cleaner.
What must you do to use it safely at home?

**10** Alkalis feel soapy. But you must never touch a chemical to see
if it might be an alkali. Why?

**11** Sodium hydroxide is very corrosive. But factories make vast
quantities of it. Why?

## Discovering acids and alkalis

Jabir ibn Hayyan was a very important
scientist. He worked in Iraq around the
year 800. Jabir discovered how to make
sulfuric acid, nitric acid and hydrochloric
acid. He was the first person to use the
word 'alkali'. Jabir wrote more than 100
books about his findings, which scientists
found useful for hundreds of years.

### Summing up

**12** Name **three** acids you might find in your body.

**13** Name **two** useful alkaline substances.

**14** Citric acid is a weak acid. Hydrochloric acid is a strong acid.
Which acid would cause less damage if you spilt it on your hand?

**Get this**

- Acids and alkalis have many uses.
- Some acids and alkalis are weak; some are strong.
- Strong acids and alkalis are hazardous.

# The pH scale

## The pH scale

It's useful to know how strong an acid or alkali is. After all, you don't want to put nitric acid on your chips instead of vinegar!

The pH scale shows how **acidic** or **alkaline** a solution is:

- The pH of an acid is less than 7. The lower the pH, the stronger the acid.
- The pH of an alkali is more than 7. The higher the pH, the stronger the alkali.
- Some solutions are neither acidic nor alkaline. They are **neutral**. Their pH is 7.

**1** Which **two** acids on this scale have the lowest pH? Which **two** alkalis on this scale have the highest pH?

**2** Name the most acidic substance shown here.

**3** Name the most alkaline substance shown here.

**4** Name the weakest acid shown here.

**5** Name the weakest alkali shown here.

**6** Name **one** neutral substance.

**7** Name **one** substance that is more alkaline than saliva.

You can find the pH of something by using **universal indicator (UI)**. When you add it to a solution it changes to one of the colours on the pH scale. This shows the pH of the substance.

**8** You add UI to pure water. What colour is the mixture?

**9** You add UI to lemon juice. What colour is the mixture?

**10** You add UI to sodium hydroxide solution. What colour does it turn?

## Why is pH testing important?

Your blood is slightly alkaline. Its pH is always 7.4. You would be very ill if your blood pH got higher than 7.6 or lower than 7.2.

The pH of your urine changes. It tells you a lot about what's happening elsewhere in your body:

- If your blood gets too acidic, extra acid comes out in your urine. So your urine pH goes down.
- If your blood is too alkaline, extra alkali comes out in your urine. So your urine pH goes up.

## Learn about
- The pH scale
- Measuring pH
- Concentrated and dilute acids and alkalis

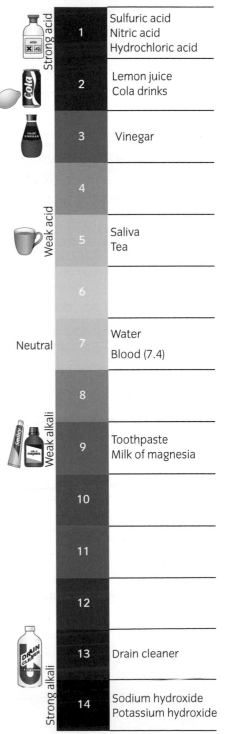

| Strong acid | 1 | Sulfuric acid / Nitric acid / Hydrochloric acid |
| | 2 | Lemon juice / Cola drinks |
| | 3 | Vinegar |
| | 4 | |
| Weak acid | 5 | Saliva / Tea |
| | 6 | |
| Neutral | 7 | Water / Blood (7.4) |
| | 8 | |
| Weak alkali | 9 | Toothpaste / Milk of magnesia |
| | 10 | |
| | 11 | |
| | 12 | |
| Strong alkali | 13 | Drain cleaner |
| | 14 | Sodium hydroxide / Potassium hydroxide |

▲ The pH scale

**11** What makes your urine pH go down when your blood is too acidic?

**12** What happens to your urine pH if your blood is too alkaline? Why?

**13** A hospital patient has a blood pH of 7.5. Explain how her body tries to get the blood pH back to normal.

## Concentrated and dilute

The lorry below is carrying ethanoic acid. This acid could give you terrible burns. But ethanoic acid – as vinegar – tastes great with chips. What's the difference?

In vinegar, ethanoic acid is mixed with lots of water. This is a **dilute** solution. But the ethanoic acid in the lorry is mixed with very little water. This is a **concentrated** solution. Concentrated acids and alkalis are more corrosive than dilute acids and alkalis.

▲ A nurse tests the pregnant woman's urine pH to check on the health of mother and baby.

**14** Which could damage your skin more – concentrated or dilute sodium hydroxide solution? Explain why.

**15** A lorry spills hydrochloric acid on the road. Fire-fighters spray water on the acid. Explain why.

## Summing up

**16** Copy and complete the table to show whether each substance is acidic, alkaline or neutral.

| Chemical | pH | Acidic, alkaline or neutral? |
|---|---|---|
| Orange juice | 3 | |
| Milk | 7 | |
| Some toothpaste | 9 | |
| Cola drinks | 2 | |
| Sweat from a teenager | 5 | |

**17** Concentrated citric acid causes serious damage to eyes. But orange squash contains citric acid. Explain why getting orange squash in your eyes is unlikely to damage them.

### Get this

- On the pH scale:
  - Under 7 is acidic
  - Over 7 is alkaline
  - 7 is neutral.
- pH testing has important uses.
- Concentrated acids and alkalis are more harmful than dilute ones.

## 5.3 Neutralisation

### What is neutralisation?

When hairdressers highlight hair they use an alkali to open the outer layer (cuticle) of each hair. This lets the dye in. Then they rinse with acid. This closes the cuticle. The acid also cancels out – or **neutralises** – left over alkali. This process is called **neutralisation**.

Bee stings are acidic. They feel better if you rub toothpaste on them. Toothpaste is alkaline. It neutralises the acidic sting.

**1** Would you use a strong acid or a weak acid to neutralise left over alkali in your hair? Give a reason for your decision.

**2** Why do hairdressers wear gloves when they highlight hair?

**3** Look at the pH scale on page 52. What could you use instead of toothpaste to neutralise a bee sting?

### Learn about
- How acids and alkalis neutralise each other
- Useful neutralisation

### How do we know... that neutralisation happens?

Charlotte had stomach ache. She took some milk of magnesia and felt better. Milk of magnesia – magnesium hydroxide – is an antacid. It is alkaline. It neutralises extra stomach acid. Charlotte did an experiment to show this.

**4** Charlotte didn't take acid from her own stomach! Use page 50 to work out which lab acid she used instead.

**5** How did the pH of the solution in the flask change during the experiment?

**6** Later, Charlotte added even more milk of magnesia to the acid. What colour did the solution become? Give a reason for your decision.

**7** How would you improve Charlotte's experiment?

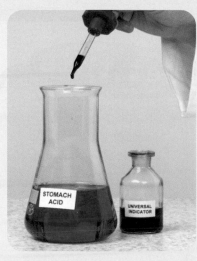

**1** The solution is acidic.

**2** The solution is less acidic. The milk of magnesia has neutralised some of the acid.

**3** The solution is neutral. The milk of magnesia has neutralised all the acid.

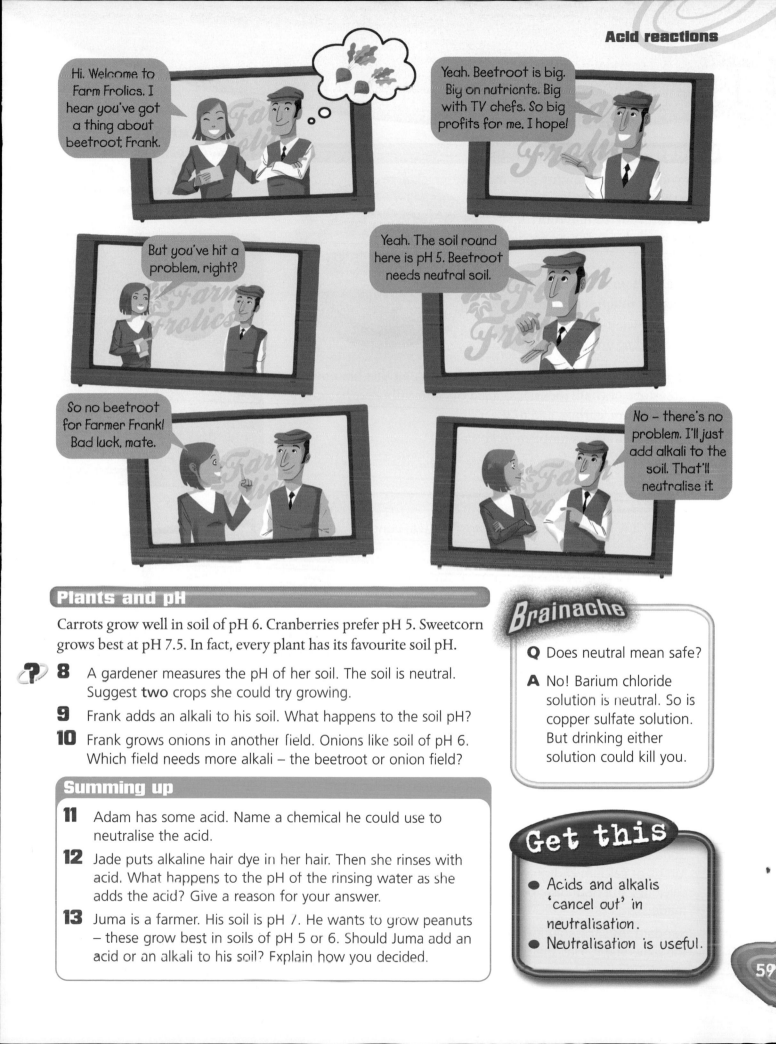

## Plants and pH

Carrots grow well in soil of pH 6. Cranberries prefer pH 5. Sweetcorn grows best at pH 7.5. In fact, every plant has its favourite soil pH.

**8** A gardener measures the pH of her soil. The soil is neutral. Suggest **two** crops she could try growing.

**9** Frank adds an alkali to his soil. What happens to the soil pH?

**10** Frank grows onions in another field. Onions like soil of pH 6. Which field needs more alkali – the beetroot or onion field?

## Summing up

**11** Adam has some acid. Name a chemical he could use to neutralise the acid.

**12** Jade puts alkaline hair dye in her hair. Then she rinses with acid. What happens to the pH of the rinsing water as she adds the acid? Give a reason for your answer.

**13** Juma is a farmer. His soil is pH 7. He wants to grow peanuts – these grow best in soils of pH 5 or 6. Should Juma add an acid or an alkali to his soil? Explain how you decided.

### Brainache

**Q** Does neutral mean safe?

**A** No! Barium chloride solution is neutral. So is copper sulfate solution. But drinking either solution could kill you.

### Get this

- Acids and alkalis 'cancel out' in neutralisation.
- Neutralisation is useful.

### Learn about
- How carbonates react with acids
- Testing for carbon dioxide

### Limescale in the loo

Does your loo have scummy grey marks under the water?
Is your kettle furred up? Your shower head clogged? Yes? You've got limescale!

Limescale is a problem. It doesn't look nice. Its rough surface means that things stick to it easily. So it's a perfect breeding ground for bacteria.

Limescale is calcium carbonate. It comes from water that has flowed over limestone or chalk rocks. It's called calcium carbonate because it contains **calcium**, **carbon** and **oxygen**.

 **1** Why is limescale a problem in kettles?

**2** Why is limescale a problem in water pipes?

### Getting rid of limescale

Shops sell many limescale removers. Most of them contain acids.

**3** Look at the label. Write the name of the acid in this limescale remover.

**4** The label shows that this limescale remover is corrosive. What precautions should you take when using it? Give a reason for each one.

**5** Is the pH of limescale remover more than 7 or less than 7? Give a reason for your choice.

Causes burns.
Keep locked up and out of reach of children.
Do not breathe vapour.
In case of contact with eyes or skin wash off immediately with plenty of cold water and seek medical help.
Wear suitable gloves and protective clothing.
In case of accident, or if you feel unwell, seek medical advice immediately (show label where possible).
Contains: Formic Acid & additives

**CORROSIVE**

### How does it work?

When you add an acid to limescale in your kettle you see fizzing bubbles. The bubbles contain carbon dioxide gas. The acid and calcium carbonate make new substances. One of the new substances is carbon dioxide.

**6** Sundara has some hydrochloric acid in a beaker. She adds lumps of calcium carbonate to it. The mixture bubbles and the calcium carbonate disappears. Name **one** new substance that is made.

**7** Sundara tests the solution with universal indicator at the end of the experiment. Has the pH gone up or down? Explain how you decided.

## What is carbon dioxide?

You can't see, smell or taste carbon dioxide gas. But it's hugely important. Plants need it to make food. Without plants, humans and other animals would starve.

Carbon dioxide in the atmosphere helps keep the Earth warm.

**8** Give **two** important uses of carbon dioxide.

## Using the fizz!

'Bath bombs' fizz in the bath. They contain two powders – citric acid and sodium hydrogen carbonate. In water, the sodium hydrogen carbonate neutralises citric acid. This makes carbon dioxide gas which fizzes into the water. Nothing happens before you put the 'bomb' in the bath. Acids need water to react.

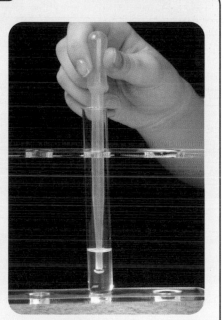

### How do we know... ...that the bubbles are carbon dioxide?

You can use limewater to find out if a gas is carbon dioxide. Limewater is colourless. It goes cloudy when carbon dioxide bubbles through it.

**9** What would you see in the limewater if the gas is carbon dioxide?

**10** What might you see if the gas is oxygen?

**11** You have two test tubes, a bendy drinking straw and some plasticine. Design a way of getting gas bubbles into limewater.

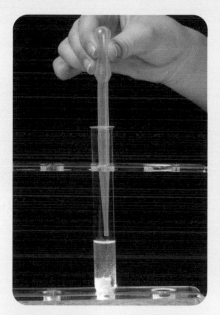

▲ Use a teat pipette to collect the gas.

▲ Bubble the gas from the pipette into limewater.

### Summing up

**12** Are limescale removers acidic, alkaline or neutral?

**13** Design an investigation to find out which acid removes limescale most quickly.
- Include these acids: vinegar, lemon juice, cola, hydrochloric acid, sulfuric acid.
- Use lumps of calcium carbonate instead of limescale.
- Decide which variables to change and which to keep the same so that the test is fair.

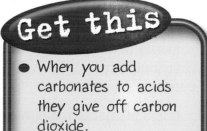

### Get this
- When you add carbonates to acids they give off carbon dioxide.
- Carbon dioxide makes limewater milky.

**Learn about**
- How acids react with metals
- Testing for hydrogen

Some coins contain zinc metal. Joel's stomach contains hydrochloric acid. The zinc and the acid **reacted** together and made two new substances: zinc chloride solution and **hydrogen** gas. Zinc chloride is poisonous. It made Joel ill.

**1** Why does Joel's stomach contain hydrochloric acid?

**2** Explain why the coin in Joel's stomach got thinner.

### Putting it to the test

You can see the same thing happening in the lab. If you put a piece of zinc metal into a test tube of hydrochloric acid, you see bubbles. After a while some of the zinc metal seems to disappear. You are left with a colourless solution.

**3** Name the gas in the bubbles.

**4** Name **one** substance in the colourless solution.

**5** What happens to the pH of the solution in the test tube? Give a reason for your answer.

## How do we know... that the bubbles are hydrogen?

Light a splint.

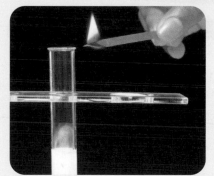

Hold the splint at the top of the test tube, just above the bubbles.

Listen! If the splint goes out with a squeaky pop, the bubbles contain hydrogen.

**6** What do you see if the gas is hydrogen?

**7** What do you hear if the gas is hydrogen?

**8** What might you see or hear if the bubbles contained carbon dioxide, not hydrogen?

## What's so special about hydrogen gas?

Hydrogen is amazing! There's more hydrogen in the Universe than anything else. Stars are mainly hydrogen. Our own star – the Sun – is about 70% hydrogen. So is planet Jupiter.

Hydrogen gas is less dense ('lighter') than air. That's why it was once used to fill airships like this one.

 **9** Name **two** things in our solar system that are made mainly of hydrogen gas.

### Brainache

**Q** Do all metals react with acids?

**A** Some, like gold and platinum, almost never do. Gold fillings don't fizz every time you eat!

## Summing up

**10** You add magnesium to an acid. You see bubbles. How could you show the bubbles contain hydrogen?

**11** Hannah is a police officer. She is examining a stolen car. The serial number which was engraved into the engine looks as if it has been scratched off. She pours acid onto the scratch marks. She sees bubbles. After a few minutes, she can read the number under the scratch marks.
  **a** What gas is in the bubbles?
  **b** The engine is made of iron. What happens to the iron when she adds acid to it?
  **c** Why must Hannah be careful not to add too much acid?

### Get this

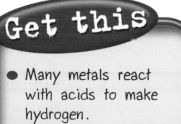

- Many metals react with acids to make hydrogen.
- In hydrogen, a lighted splint goes out with a squeaky pop.

## Neutralisation

What do antiperspirants, hair shampoo, cough medicine, batteries and fertilisers have in common? They may all contain the same important salt: ammonium chloride.

People have valued ammonium chloride for centuries. They've used it to treat chest infections, dissolve metals and dye clothes.

It wasn't always easy to get hold of ammonium chloride. Many years ago people risked their lives getting ammonium chloride from burning coal caves in Tajikistan or from Persia's 'Smoking Fountain of Hell' volcano.

There are simpler ways to get ammonium chloride! You can neutralise ammonium hydroxide (an alkali) with hydrochloric acid. This makes a solution of ammonium chloride (a salt) in water.

**1** Name the acid that neutralises ammonium hydroxide to make ammonium chloride.

Whenever an acid neutralises an alkali, two chemicals are made: a **salt** and water:

- Sulfuric acid neutralises sodium hydroxide to make sodium sulfate (a salt) and water.
- Nitric acid neutralises potassium hydroxide to make potassium nitrate (a salt) and water.

The name of the salt comes from the name of the acid:

- Hydrochloric acid makes chloride salts.
- Sulfuric acid makes sulfate salts.
- Nitric acid makes nitrate salts.

**2** What salt do you make by neutralising potassium hydroxide with sulfuric acid?

**3** Which **two** chemicals do you make by neutralising sodium hydroxide with nitric acid?

**4** What acid could you add to potassium hydroxide to make potassium chloride?

**5** What alkali could you add to hydrochloric acid to make sodium chloride?

**Learn about**
- Salts from reactions with acids
- Writing word equations

▲ Ammonium chloride

▲ Burning coal caves in Tajikistan.

## Acids and carbonates

When you add hydrochloric acid to limescale – calcium carbonate – you see bubbles of carbon dioxide gas. You also make two other new chemicals: calcium chloride (a salt) and water.

This **word equation** describes what happens:

hydrochloric acid + calcium carbonate $\rightarrow$ calcium chloride + carbon dioxide + water

The equation shows that you start with two chemicals – citric acid and calcium carbonate. You make three new chemicals – calcium chloride, carbon dioxide and water.

You can write similar word equations for the reactions of other carbonates with acids. For example:

sulfuric acid + magnesium carbonate $\rightarrow$ magnesium sulfate + carbon dioxide + water

**6** You add sulfuric acid to magnesium carbonate. Name the three new chemicals that are made.

**7** You add nitric acid to sodium carbonate. Copy and complete the word equation to show the three new chemicals that are made.

nitric acid + sodium carbonate $\rightarrow$ _____ + _____ + _____

**8** You add hydrochloric acid to barium carbonate. Write a word equation to represent what happens.

**9** You add sulfuric acid to sodium carbonate. Write a word equation to represent what happens.

## Acids and metals

If you drop a small piece of magnesium metal into a test tube of hydrochloric acid you see bubbles of hydrogen gas. You also make a salt: magnesium chloride. You can't see the magnesium chloride because it is dissolved in water.

This word equation describes what happens:

magnesium + hydrochloric acid $\rightarrow$ magnesium chloride + hydrogen

**10** You add zinc to sulfuric acid. Name the salt that is made.

**11** You add lead to nitric acid. Name the two chemicals that are made.

**12** You add magnesium to nitric acid. You make magnesium nitrate and hydrogen. Write a word equation to represent what happens.

**13** You make zinc chloride by adding zinc to an acid. Write a word equation for this. Include the name of the acid you used.

## Summing up

**14** Copy and finish the equations below:

acid + alkali $\rightarrow$ salt + _____

acid + carbonate $\rightarrow$ salt + _____ + _____

**Brainache**

**Q** What does the arrow in a word equation mean?

**A** The arrow means 'makes'. It is not the same as an = sign in a maths equation!

**Get this**

- An acid and an alkali make a salt and water.
- An acid and a carbonate make a salt, carbon dioxide and water.
- An acid and a metal make a salt and hydrogen.
- Word equations show what happens in reactions.

Did you wash yesterday? Clean your teeth? Make yourself a drink? Scientists make sure your water is bug-free. They check its pH too. After all, no one wants water so acidic that it dissolves the pipes...or so alkaline that it blisters their skin.

This canal is really important. Half a million people rely on it to bring them water from the River Severn. The water supply company tests the canal water in four places along the canal. It uses pH meters to measure the pH. The pH meters are linked to a computer system. A sudden pH change shows there might be a problem. This sets off a central alarm, even in the middle of the night. Then it's down to the duty scientist to try to solve the problem – and fast!

### Learn about

- Why water pH matters to humans and animals

**1** A boat accidentally spills hydrochloric acid into the canal. What happens to the pH of the water?

**2** What type of chemical might the company add to the water to get the pH back to normal?

## Testing the water

Biologists and chemists have found the best way of treating the canal water to make it safe to drink. The process involves several pH changes. The diagram on the right shows where these changes happen.

**3** For each of the stages *a* to *d*, work out whether the water company adds an acid or an alkali or neither to the water.

Patric Bulmer plays a vital role in getting safe water to a million people. He's worked at the water company since getting a degree in geochemistry a few years ago.

Take water from the lake.
**Its pH is between 7.7 and 8.9.**

*a*

Add aluminium chloride to make the water clear.
**Works best at pH 6.8.**

Filter through sand.

*b*

Bubble ozone through the water to destroy bacteria, viruses and pesticides.
**Works well between pH 6.0 and 7.0.**

*c*

Filter through carbon.
**Water pH does not change.**

Add chlorine to destroy bacteria that get into the water in the pipes.
**This makes the water more acidic.**

*d*

Send water in pipes to people's homes.
**Water pH is around 7.5.**

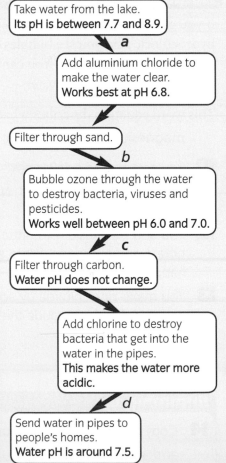

◄ Patric and his colleague Vicky Richards testing the water.

Patric says, 'I love the variety in my job. It's hands-on, too. The other day I was clambering around in a water tower checking for bugs. Then a customer phoned to discuss the taste of water at his home. Later I made new dipping rods to get water samples from the lakes. The next day I was discussing climate change – and what we can do about it – with other scientists in London.'

## pH for life

Water companies are not the only organisations that care about water pH. Environment and conservation organisations do too. Chemists have collected data about lake pH. Biologists have discovered the water pH at which different animals and plants survive.

| Name of lake | pH of lake water in | | | |
|---|---|---|---|---|
| | 1960 | 1970 | 1990 | 2001 |
| Llyn Llaghi, Wales | – | – | 5.0 | 6.0 |
| Lake Gårdsjön, Sweden | 6.0 | 4.5 | – | – |

**4**
**a** What was the pH of Lake Gårdsjön in 1960?

**b** What happened to the acidity of Lake Gårdsjön between 1960 and 1970?

**c** Name **two** animals that might have lived in Lake Gårdsjön in 1960.

| Animal or plant | Water pH the animal or plant can live in | | | | | | | |
|---|---|---|---|---|---|---|---|---|
| | 4.0 | 4.5 | 5.0 | 5.5 | 6.0 | 6.5 | 7.0 | 7.5 |
| Trout | | | ✓ | ✓ | ✓ | ✓ | ✓ | ✓ |
| Salmon | | | | ✓ | ✓ | ✓ | ✓ | ✓ |
| | | ✓ | ✓ | ✓ | ✓ | ✓ | ✓ | ✓ |
| Eel | | | | | ✓ | ✓ | ✓ | ✓ |
| Snails | | | | ✓ | ✓ | ✓ | ✓ | ✓ |
| Mayfly | | | | | ✓ | ✓ | ✓ | ✓ |

**d** Name **one** animal that might have lived in Lake Gårdsjön in 1970.

**e** A biologist found lots of snails in Lake Gårdsjön in 2007. What might have happened to the lake's pH to make this possible?

**5**
**a** A biologist found no mayflies in Llyn Llaghi in 1990. Suggest **one** possible reason for this.

**b** The biologist looked for mayflies in Llyn Llaghi in 2001. Do you think she found any? Give a reason for your answer.

## Get this

- Water companies monitor the pH of water sources and treat drinking water.
- The pH of water affects the animals and plants that can live in it.

**Learn about**
- Particles in solids, liquids and gases
- Scientific evidence and theories

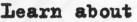

## What is matter?

You buy a bar of chocolate. How many pieces can you divide it into? Enough to share with your best friend? Everyone at school? Everyone in the world?

Greek philosophers wondered about a question like this over 2500 years ago. They wanted to know if you could go on cutting a piece of stuff – **matter** – into smaller and smaller pieces for ever.

Zeno answered yes. He said that matter completely fills its space. So you can keep cutting it into smaller pieces for ever.

Leucippus and Democritus thought differently. They said that matter is divided into tiny separate bits – **particles** – with empty space between. They believed that particles are the smallest pieces of matter you can have. This was their **theory**.

Zeno, Leucippus and Democritus did not do practical experiments. They worked out their answers by thinking carefully and creatively about the problem.

 **1** Do your own thought experiment. Imagine cutting a bar of chocolate into smaller and smaller pieces. Could you go on cutting for ever? Is it possible to decide who was right – Zeno, or Leucippus and Democritus?

## The particle theory

The argument continued for centuries until the early 1800s. Then John Dalton, a Manchester science teacher, experimented on gases. He also looked at results from other scientists' experiments. He thought creatively about all this **evidence**. Eventually he came up with his **particle theory**.

Gradually scientists collected more evidence to support the theory. They found that they could use it to correctly **predict** how matter behaves. Now all scientists accept that everything is made of particles.

**2** What evidence did Dalton use to persuade other scientists to believe his theory?

## Solids, liquids and gases

The particle theory explains the behaviour of solids, liquids and gases.

Why can't I change the shape of this rock?

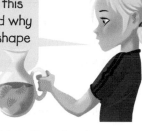

How come I can pour this drink? And why does its shape change?

This gas is going everywhere – and it doesn't have fixed shape! Why?

It's a **solid**. In solids, the particles are in a regular pattern, touching each other. They are strongly attracted to each other. The particles don't move around – but just vibrate on the spot.

It's a **liquid**. In liquids, the particles touch each other, but they're not arranged in a regular pattern. They are still strongly attracted to each other. The particles move around, in and out of the other particles.

In a **gas** the particles don't touch each other. They're not attracted to each other either. The particles move fast in all directions – there's no regular pattern.

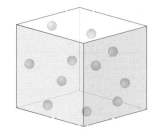

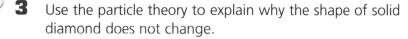

**3** Use the particle theory to explain why the shape of solid diamond does not change.

**4** Use the particle theory to explain why apple juice takes the shape of the glass it's in.

### Summing up

**5** Copy and complete the sentences using the words below.

evidence    experiments    predictions    theories

Scientists do _____ and make observations to collect _____. They think creatively about evidence and come up with _____. Theories help to make _____.

**6** Copy and complete this table:

|  | How close are the particles? | Are they in a pattern? | How do they move? | How strongly do they attract each other? |
|---|---|---|---|---|
| Solids |  |  |  |  |
| Liquids |  |  |  |  |
| Gases |  |  |  |  |

### Get this

- Matter is made of tiny particles.
- The particle theory explains how solids, liquids and gases behave.
- Scientific theories are based on evidence.
- Scientists use theories to explain observations and make predictions.

### Squashing solids, liquids and gases

There's enough oxygen in this cylinder to support a climber for 20 hours at the top of Mount Everest. Here, the gas is squashed into one small cylinder. Normally, this amount of oxygen fills 2270 empty drinks cans – imagine trying to carry that!

You can squash – **compress** – all gases. When you do, the particles get closer together.

These drawings show the particles in two dimensions. Three dimensional drawings – like those on page 69 – make the particles easier to imagine, but are harder to draw.

You can't compress solids or liquids. Their particles are already touching so they can't get closer together.

**1** Which of these can you compress: nitrogen gas, orange juice, sand?

**2** You can compress gases, but not solids or liquids. How does this evidence support the particle theory?

**3** Your camping stove has a leak. Where do the gas particles go?

### Spreading out

Daisy wears new perfume. Very soon others notice it. Why?

Perfume particles evaporate from Daisy's skin. They move around in the air. So they spread out and mix with moving air particles. Soon there are perfume particles all over the room. This 'spreading out' is **diffusion**. Some of the perfume particles enter your nose, which detects the smell.

**4** What is diffusion?

**5** Are the perfume particles that enter your nose in a solid, liquid or gas?

**6** Air is a mixture of gas particles. What do the gas particles do?

### Diffusing liquids

Many swimming pool owners add liquids containing chlorine to make their pools safe. Particles from the liquid spread out – **diffuse** – through the whole pool.

**7** Liquid diffusion is slower than gas diffusion. Draw and label **two** particle diagrams to explain why.

- Squashing solids, liquids and gases
- Particles spreading out

▲ When you compress a gas... the particles get closer together.

## Can solids diffuse?

Solids diffuse into other solids. This can cause problems. Solar cells generate electricity from light. This solar cell has a light-absorbing film on a sheet of metal. The cell is damaged if metal particles diffuse into the film. Scientists prevent diffusion by putting aluminium oxide between the film and the metal.

light-absorbing film
aluminium oxide
metal base

**8** Solid diffusion is slower than liquid diffusion. Use ideas about particles to explain why.

## Scientific models

The particle theory is a model. Scientific models help you imagine things you can't see. You can use models to:

- explain things that happen
- make predictions.

### How do we picture... the particle model?

Use the diagrams on page 69… or take a box of maltesers:

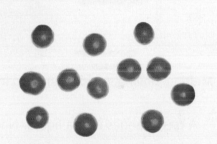

Solid: the maltesers have a regular pattern. They touch their neighbours.

Liquid: the maltesers touch their neighbours. They move around each other.

Gas: the maltesers move fast in all directions. They don't touch each other.

**9** Give **two** ways in which the maltesers model is helpful in picturing particles in solids, liquids and gases.

**10** In what ways is the model not helpful?

## Summing up

**11** Copy and complete the table.

|  | Solids | Liquids | Gases |
|---|---|---|---|
| How quickly do they diffuse? |  |  |  |
| How squashy are they? |  |  |  |

### Get this

- You can compress gases but not liquids or solids.
- In diffusion particles spread out.
- Gases diffuse fast; solids diffuse slowly.

**Learn about**
- Changes of state
- Expansion and contraction

## States of matter

Laura lights a candle. Some of the solid wax becomes liquid. Some wax changes into a gas, which burns.

Like wax, most substances can exist as a solid, a liquid or a gas. These are the **states of matter**.

**1** What states of matter do pictures 1 and 2 show?

**2** What change does the arrow represent?

**3** What is the name for this change?

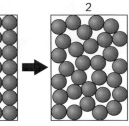

## Changing state

When a substance changes state, its particles don't change. All that changes is the distance between the particles, their speed and the attraction between them.

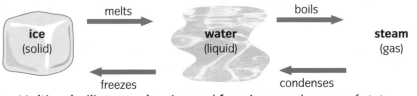

▲ Melting, **boiling**, **condensing** and **freezing** are changes of state.

**4** A gas changes into a liquid. Name this change of state.

**5** A liquid freezes. What is its new state?

## Getting bigger and smaller

George tries to open a jar. Its metal lid won't budge. So he runs hot water over the lid. It comes off easily. This works because the lid gets slightly bigger – **expands** – as it gets hotter and so fits less tightly.

Liquids also expand. In this thermometer, liquid alcohol expands as it gets hotter. So it takes up more space in the tube. The alcohol level goes up and the scale shows the temperature.

**6** You heat solid silver. How does it change before it gets hot enough to melt?

Solids and liquids expand as they get hotter because their particles vibrate faster, so they get slightly further apart.

When a solid or liquid cools its particles get closer together. So the substance gets smaller, or **contracts**.

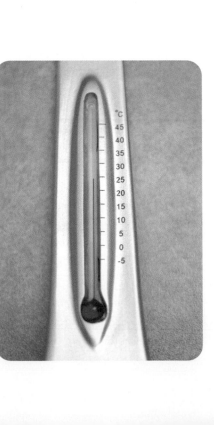

**How do we show...** that gases also expand and contract?

▲ In the freezer the air in the bottle gets cold. It contracts.

▲ Under the hot tap the air warms up. It expands.

**7** Draw particle pictures to show the air particles in the warm and cold bottles.

## Melting and boiling points

Every pure substance changes state at particular temperatures. Ice melts at 0 °C. Its **melting point** is 0 °C. The **boiling point** of water is 100 °C. Chocolate melts at about 36 °C, just below body temperature.

**8** You put solid chocolate in your mouth. What happens? Why?

Some substances have very high melting and boiling points – much hotter than the hottest British day ever, of over 38 °C.

Look at the table.

▲ Liquid gold.

| Substance | Melting point (°C) | Boiling point (°C) |
|-----------|--------------------|--------------------|
| Aluminium | 660 | 2 470 |
| Gold | 1 063 | 2 970 |

**9** In what states are gold and aluminium at 2600 °C?

**10** What change of state happens when a scientist cools aluminium from 3000 °C to 2000 °C?

Other substances change state at very low temperatures. Methane melts at –182 °C and boils at –164 °C (much colder than a freezer at –18 °C!).

**Brainache**

**Q** Do particles get bigger when they get hotter?
**A** No. The particles stay the same size. They just get further apart.

**Get this**

- Substances change state at particular temperatures.
- Substances expand on heating and contract on cooling.

## Summing up

**11** Liquid nitrogen boils. What is its new state?

**12** Draw **two** particle pictures linked by an arrow to show condensing.

73

## Learn about
- Speeding up dissolving
- Finding the masses of solutions
- Saturated solutions

## Speeding up dissolving

Marco makes iced coffee. He adds coffee powder to a little hot water and stirs. Then he fills the glass with cold milk and ice.

When Marco adds coffee powder to hot water, the powder **dissolves**. Marco makes a **solution**. Water is the **solvent**. Coffee powder is the **solute**.

**1** Marco uses hot water to make the coffee powder dissolve quicker. How else does he speed up dissolving?

**2** Lalita adds sugar to cold water. She stirs. The sugar dissolves. What are the solvent and the solute?

**3** What is the scientific word for the sugar and water mixture?

**4** How can Lalita make the sugar dissolve quicker?

### How do we picture... the particles in solution?

▲ Particles in solid sugar        ▲ Particles in liquid water.        ▲ Particles in sugar solution.

You can represent particles in a solution with rice and dried beans. The rice represents water particles. The beans represent sugar particles.

**5** In what ways is the rice and bean model good?

**6** How is this model *not* like the particles in sugar solution?

### How do we know the particles are still there?

You dissolve sugar in a glass of water. The sugar seems to disappear. How can you tell that the sugar is still there (without drinking it)?

- Weigh the water.
- Weigh the sugar.
- Add the sugar to the water and weigh the solution.

The mass of the solution is the same as the masses of the water and sugar added together.

**8** Abdul dissolves 5 g of salt in 200 g of water. What mass of solution does he make?

**9** A company makes a medicine to relieve pain and fever. It dissolves 24 g of Paracetamol and 2.5 g of diphenhydramine in 973.5 g of water. What is the mass of the medicine?

**10** The medicine company uses warm water and stirs the mixture. Why?

## How much solid can dissolve in a liquid?

Luke likes sweet tea. One day, for an experiment, he puts 60 spoons of sugar in a mug of tea. Some sugar stays at the bottom. It doesn't dissolve. Luke has made a **saturated solution**.

There is a limit to the amount of solid that dissolves in 100 g of water. The limit is different for different solids.

The bar chart shows the maximum amounts of four solids that will dissolve in 100 g of water at 20 °C. It shows how **soluble** the solids are.

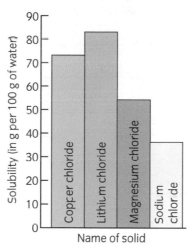

**11** Which solid is most soluble?

**12** Which solid is least soluble?

**13** Which solid makes a saturated solution when you add 40g of it to 100 g of water?

### Summing up

Olivia dissolves 2 g of copper sulfate in 100 g of water.

**14** Name the solvent and the solute.

**15** What is the mass of the solution?

**16** What happens to copper sulfate particles when they dissolve?

Olivia adds another 19 g of copper sulfate to the solution. Not all of it dissolves.

**17** What type of solution has she made?

*Get this*

- Stirring and heating speed up dissolving.
- In saturated solutions, no more solute can dissolve.
- You can find the mass of a solution by adding the masses of solvent and solute.

## Gas pressure

Raj blows up a balloon. He pumps harder and harder. The balloon gets bigger and bigger. Suddenly – bang – it bursts! Why?

Raj begins to pump air particles into the balloon. The air particles move in all directions. They bump into – **collide** with – the rubber. The collisions push the rubber outwards. This is **air pressure**.

More air particles enter the balloon making more collisions with the rubber. The pressure inside the balloon increases.

Air particles hit the outside of the balloon too. When more particles hit the inside of the balloon than the outside, the pressure inside is greater. If the air pressure inside gets really high, the balloon bursts.

 **1** The gas pressure inside a bag of crisps is slightly greater than the pressure outside. This stops the crisps getting crushed. Draw a picture to show gas particles inside and outside the bag.

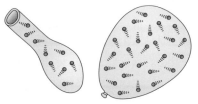
▲ Blowing up a balloon.

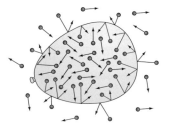
▲ Air particles hit both the inside and the outside of the balloon.

## Using gas pressure

Pressure chambers can help heal footballing injuries. A specially strengthened pressure chamber contains pure oxygen gas, and the pressure inside it is twice normal air pressure. So the footballer breathes in more oxygen particles than normal. This speeds his recovery.

**2** Are there more collisions on the inside or outside of the pressure chamber walls? Give a reason for your decision.

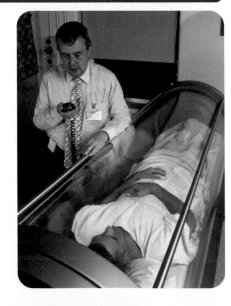

## How does temperature affect gas pressure?

When air particles get hotter, they move faster. The air in this bottle is cold. So the particles move slowly and hit the inside of the bottle less often. Every minute, more air particles hit the outside of the bottle than hit the inside. So the air pressure outside is greater, and the bottle collapses.

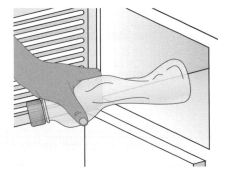

The air particles in this bottle move faster. They hit the inside of the bottle more often. More air particles hit the inside of the bottle than the outside. So the pressure inside is more than the pressure outside.

**3** The lid blows off this container when it gets too hot. Why?

## Density

**4** Look at the pictures. Which is heavier – steam or liquid water?

**5** Which bottle do you think contains more particles?

The bottle of liquid water is heavier. The particles are more closely packed together, so more of them fit in the bottle. Liquid water has a greater **density** than steam.

Density is how heavy something is for its size. Solids are usually more dense than liquids and gases. Gases have the lowest densities.

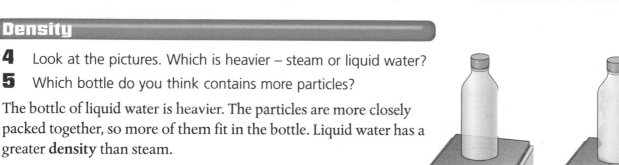

▲ Liquid water.     ▲ Steam.

## How do we know... which is more dense – a solid or a liquid?

Find the mass of 1 cm³ of the solid. Then find the mass of 1 cm³ of the liquid. The one with the greater mass has the higher density.

**6** Which is heavier – 1 cm³ of solid or liquid gold?

**7** Which has the higher density?

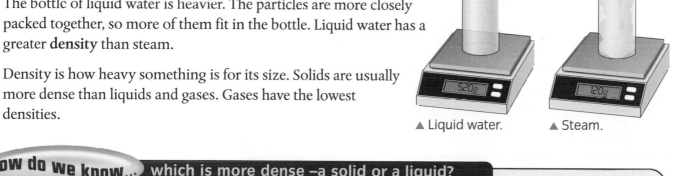

## Summing up

**8** The pressure inside a tyre is more than the air pressure outside. Use the particle model to explain why.

**9** The pressure inside a tyre goes up on hot days. Explain why.

**10** Describe an experiment you could do to find out which is more dense: sugar, salt or sand.

## Get this

- Gas pressure is caused by particles colliding with container walls.
- Heating increases gas pressure.
- Most solids are denser than liquids and gases.

If you lived in Eritrea or Ethiopia you'd probably spend a lot of time at coffee ceremonies. These are a great chance to catch up on gossip, discuss politics or simply relax over three delicious cups of coffee. The coffee is definitely not instant – the ceremonies may take hours!

Fekerte roasts the coffee beans over a charcoal stove. The guests enjoy the scent of the roasting beans, mingled with the smell of burning incense.

**? 1** Roasting coffee beans produce beautifully scented (aromatic) gas particles. The particles spread out. What do scientists call this spreading out?

**2** Draw a diagram to show the particles as they start to spread out.

**3** Explain scientifically why Fekerte smells the aromatic particles before her guests smell them.

Fekerte grinds the coffee beans in a pestle and mortar (or whizzes them in a food processor) to make a powder. She then adds the coffee powder to water in a special coffee pot. She shakes the pot and puts it on the stove.

**? 4** Fekerte wants the coffee to dissolve in the water. What **two** things does she do to speed up the dissolving?

The mixture of coffee and water gets hotter. After a while, some of it bubbles up through the neck of the pot. The guests also see steam.

**5** When the water boils, the coffee pot contains water in two states – give their names.

**6** Draw diagrams to show how the particles are arranged in each of these states.

**7** There is no lid on the coffee pot. What might happen if there was a lid? Use ideas about particles to explain why this might happen.

Fekerte takes the pot off the heat. She pours coffee into tiny china cups, and adds salt or sugar. She stirs the drink in each cup.

**8** The drink is a solution.
  **a** Name the solvent.
  **b** Name **two** of the solutes.
  **c** Draw a diagram to show how the particles of solvent and solutes are arranged in the cup of coffee.

**9** Why does Fekerte stir the drinks in the cups?

The guests enjoy the strong brew!

Fekerte adds more water to the ground coffee in the pot. She heats it again until the mixture boils. Then she pours everyone another cup of coffee. This tastes less strong than the first cup.

Fekerte adds more water to the ground coffee in the pot. She heats it for a third time until the mixture boils. Then she pours each guest a final cup of coffee. This tastes much weaker than the first cup – but it's still delicious!

**10** This diagram represents the particles in the first cup of coffee. Draw another diagram to show the particles in the second, less strong, cup of coffee.

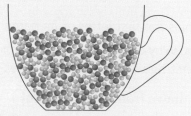

**11** Which drink contains the more dilute solution – the first or second cup? Use page 57 to help you.

**Get this**

- The particle theory explains the states of matter, diffusion and dissolving.

## What's everything made of?

Look around you. How many **materials** can you see? There are millions of different materials. They do an amazing variety of things.

How we use a material depends on its **properties** – what the material is like. So, wood is strong and it looks nice. These two properties mean that wood is a good material to make chairs from.

 **1** What property of glass means that it makes good windows? What property of glass means that it's not a *perfect* material for making windows?

Every material is made from one or more **elements**. There are about 100 elements. You can't split an element into anything simpler.

The pictures show some elements – and how we use them. Of course, elements are materials too!

**Brainache**

**Q** Are fire and water elements?

**A** No! But 2300 years ago the Greek thinker Aristotle thought that everything was made from combinations of fire, water, air and earth.

▲ Gold makes stunning jewellery.

▲ Sulfur is used to make gunpowder for fireworks and guns.

▲ These balloons are filled with helium gas.

## Symbols for the elements

Some elements have long names. So each element has its own **symbol**. It's much easier to write 'Pr' than 'praseodymium'!

Often, the symbol is the first one or two letters of an element's name in English.

**2** Name an element whose symbol is the first letter of its name.

**3** Look at the symbols in the table. Why do you think the symbols of calcium and helium are the first two letters of their names?

| Symbol | Element |
|--------|---------|
| H | hydrogen |
| He | helium |
| C | carbon |
| N | nitrogen |
| O | oxygen |
| P | phosphorus |
| Ca | calcium |

Sometimes symbols for elements are drawn from their names in other languages.

**4** Why do you think that iron has the symbol Fe?

**5** From what language does tungsten get its symbol?

| Symbol | Name of element ... | | |
|--------|---------|---------|---------|
| | **in English** | **in Latin** | **in German** |
| **Fe** | iron | ferrum | Eisen |
| **W** | tungsten | – | Wolfram |

Scientists all over the world use the same symbols. All scientists recognise S as the symbol for sulfur – even though it's called *gundhuk* in Hindi, *iwo* in Japanese, *soufre* in French and *azufre* in Spanish!

**? 6** What symbol do Chinese scientists use for the element carbon?

## Inside elements

Elements are made of tiny bits – particles – called **atoms**. Each element has its own type of atom.

**How do we picture...** whats inside elements?

Toy bricks are a useful model.

▲ If these bricks represent oxygen atoms...

▲ ... then these represent atoms of another element – say nitrogen.

▲ If these bricks represent gold atoms...

▲ ... then this shows a piece of gold.

Of course, the model is not perfect. For one thing, atoms don't have straight edges. And atoms are much, much, much smaller than the toy bricks!

## Summing up

**7** What is an element?

**8** Give the names and symbols of **five** elements.

**9** What is an atom?

**10** Are the atoms in a piece of gold the same as, or different from, each other?

**11** Are oxygen atoms the same as, or different from, gold atoms?

## Get this

- Everything is made from elements.
- You can't split an element into anything simpler.
- Each element is made from its own type of atom.

**Learn about**
- Metals
- Non-metals

### Metals

Do you use a phone? Ride a bike? Wear jewellery? All these contain **metals**. Most elements are metals. They are incredibly useful.

**?** **1**　Name **five** things you use that contain metals.

Metal elements have similar properties. Their properties explain how we use them.

Most metals are solid at room temperature.
Most metals are strong. That's why it's difficult to break or squash them.

Many metals are hard. It's not easy to cut or scratch them. Thin metal sheets, and metal wires, are bendy. So when a car crashes its metal body doesn't break into lots of little pieces – it just bends.

Heat and electricity travel quickly through metals. Some metals conduct electricity better than others – the best are copper and silver.

Metals are shiny when you first cut them, or if you polish them. After a while, most metals go dull on the outside. But gold and platinum are always shiny – that's why they make such good jewellery.

**?** **2**　Give **three** properties of a typical metal.

**3**　Why do you think that central heating radiators are made of metal?

**4**　Give **two** properties of gold that explain why it makes good jewellery.

### Strange metals

Not all metals are solid, strong, hard, bendy and shiny. One metal, mercury, is liquid at room temperature. A few metals, like potassium and sodium, are as soft as butter – you can easily cut them with a knife. Calcium is also quite soft. Potassium, sodium and calcium are all silver-coloured when you cut them, but they quickly get a white coating in air.

**?** **5**　How is mercury different from other metals?

**6**　How are calcium, sodium and potassium different from other metals?

## Non-metals

Only about 20 elements are not metals. They are called **non-metals**. Many are gases at room temperature. Nitrogen and oxygen make up most of the air. Helium fills balloons and neon makes signs glow brightly. The gases hydrogen and chlorine are also non-metal elements.

A few non-metals – like sulfur, silicon and phosphorus – are solid at room temperature. Most of them are not shiny. They usually break easily if you hit them with a hammer. You can't bend solid non-metals. Most non-metals don't conduct electricity.

Carbon is a very important non-metal element. There are several types of solid carbon. In each type, the atoms have a different arrangement …

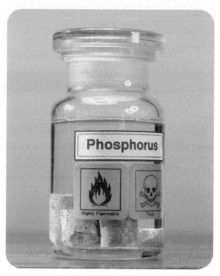

▲ 'White' phosphorus.

▲ This is charcoal.

▲ This is graphite…

▲ …and this is diamond.

**7** Name **two** non-metals that are gases at room temperature.

**8** Give **two** properties of sulfur.

## The elements of life

Every living thing is made mostly from non-metals. Your muscles are mainly carbon, hydrogen, oxygen, nitrogen and sulfur. The picture shows the elements in the body of a 50 kg person.

**9** Which **three** elements does you body contain most of?

**10** What percentage of your body mass consists of non-metals?

32.5 kg Oxygen

9 kg Carbon

5 kg Hydrogen     1.5 kg Nitrogen

1.5 kg metals

0.5 kg Phosphorous

## Summing up

**11** Copy and complete the table.

|  | What they look like | Do they usually conduct electricity? | Can you bend them? |
|---|---|---|---|
| **Metals** |  |  |  |
| **Non-metals** |  |  |  |

### Get this

- Metals and non-metals have different properties.
- Living things contain mainly non-metals.

# Compounds

## Tom's teeth

Tom goes to the dentist. He's happy – he needs no fillings! The dentist tells Tom to drink milk to keep his teeth healthy. Milk contains calcium, just like tooth enamel.

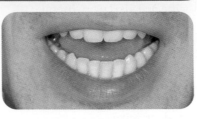

**Learn about**
- Compounds
- Differences between compounds and elements

Calcium is an element. It's shiny when you cut it and fizzes if you put it in water. Tooth enamel contains other elements, too:

- phosphorus – a poisonous white solid that catches fire easily, making clouds of deadly white smoke
- oxygen – a gas that helps things burn.

So why don't your teeth catch fire? Or poison you? Or fizz when you drink water?

## Compounds

Tooth enamel elements are not just mixed up. Their atoms have joined together to make one material – calcium phosphate. This material is totally different from all the elements in it.

Like tooth enamel, most materials are not just elements on their own. They are made of atoms of elements joined to atoms of other elements. These materials are **compounds**.

▲ Calcium in water.

**1** Name **one** metal element in tooth enamel.

**2** Name **two** non-metal elements in tooth enamel.

**3** What is a compound?

**4** Name **two** elements in the compound calcium phosphate.

## Joining up makes a difference

The properties of a compound are totally different from the properties of the elements in it.

Sodium is a shiny metal. It fizzes in water. Chlorine is a green, smelly, poisonous gas. These two elements join together to make a compound – sodium chloride – the salt some people sprinkle on food!

▲ Burning white phosphorous.

**5** Give **one** property of sodium, chlorine and sodium chloride.

**6** Are the properties of sodium chloride similar to, or different from, those of the elements in it?

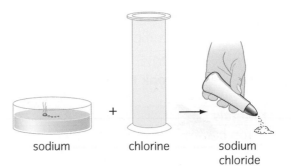

sodium + chlorine → sodium chloride

Carbon is a solid. Oxygen is a colourless gas – you can't live without it. Join the right amounts of these elements together and you make the compound carbon monoxide. This gas is a deadly poison.

The names of many compounds show which elements are in them. Usually the name of one element is changed slightly.

**7** How is the name 'sodium chloride' different from the names of the elements in it?

## How do we picture... what's inside compounds?

At room temperature, carbon monoxide particles are far apart. They move around. So carbon monoxide is a gas.

▲ You can use these bricks to represent carbon atoms...

▲ ...and these to represent oxygen atoms.

▲ You can join the bricks like this to show the atoms in carbon monoxide (a compound). Each carbon monoxide particle contains one carbon atom and one oxygen atom.

It's not always this easy to represent particles in compounds!

**8** How many carbon atoms are in one carbon monoxide particle?

**9** How many oxygen atoms are in one carbon monoxide particle?

**10** What is the total number of atoms in one carbon monoxide particle?

**11** Give **one** property of carbon and **one** property of carbon monoxide.

### Summing up

**12** Give **one** difference between elements and compounds.

**13** Hydrogen and oxygen are colourless gases. They join together to make water. Give **one** property of water which is different from the properties of the elements in it.

Jake, Kate and Magda are at a beach barbeque. They're having problems…

*Yuk! There's sand in my water. How can I get rid of it?*

*There's sand in the strawberries too. What can we do about that?*

## Learn about
● Useful mixtures
● Differences between mixtures and compounds

*We forgot the salt! Can we get some from the sea?*

Sand and water, sand and strawberries, and salty water, are all **mixtures**. The substances in mixtures are not joined to each other. They're just mixed up.

It's usually quite easy to separate the substances in a mixture from each other.

**1** Jake, Kate and Magda have this equipment. How can they use it to get clean water from the sandy water, clean strawberries and salt from seawater?

**2** Give **one** difference between mixtures and compounds (use page 84 to help).

## Different mixtures

Mixtures can contain elements, compounds or both.

Salim has two sorts of nails – iron nails and copper tacks. He keeps them in a jar. Iron and copper are both elements. So the jar contains a mixture of elements.

The label shows the ingredients in the toothpaste. The ingredients are all compounds. So toothpaste is a mixture of compounds.

Some mixtures contain both elements and compounds. Air is a mixtures of gases, including:

● elements, like nitrogen, oxygen and argon
● compounds, like carbon dioxide.

**3** Suggest how you could separate the iron nails from the copper tacks.

**4** Name **three** elements and **one** compound in air.

**SMILES**

INGREDIENTS
Aqua, Hydrated Silica, Sorbitol, Glycerin, PEG-6, Sodium Lauryl Sulfate, Aroma, Titanium Dioxide, Xanthan Gum, Carrageenan, Sodium Fluoride, Sodium Saccharin, Limonene.

## Properties of mixtures

The properties of a mixture are similar to the properties of the substances in it.

The label shows some of the compounds in orange squash.

Freshly squeezed orange juice is a mixture of compounds, too.

**CONTAINS**
water, glucose, fructose, sucrose, citric acid, sodium benzoate

 **5** Name **three** compounds that are in both orange squash and orange juice.

Glucose, fructose and sucrose are sugars. They make things taste sweet. Citric acid and ascorbic acid have sour tastes. The mixture of these, and other, compounds gives orange juice its flavour.

Taste is a property. The taste of orange juice (a mixture) is similar to the tastes of the sugars, acids and other compounds that are in it.

**6** Why does orange juice taste sweet, and slightly sour (sharp) at the same time?

## Amounts in mixtures

Often, the amounts of the substances in a mixture don't matter very much. You can add a lot or a little orange squash to a drink – but either way it tastes of orange squash.

Sometimes the amounts do matter. These tablets contain a mixture of substances:

- active ingredients to make you feel better (2,4-dichlorobenzyl alcohol and amylmetacresol)

- other compounds to make them taste nice (sucrose, glucose syrup, tartaric acid and peppermint oil).

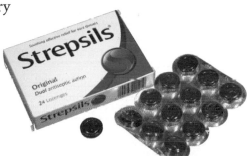

**7** Imagine there was too little amylmetacresol in one of the tablets. What problem might this cause?

**8** One tablet contains 1.5 g of sucrose. Would a tablet containing only 1.0 g of sucrose help you feel better? Explain how you decided.

### Summing up

**9** Copy and complete the table to show **two** differences between mixtures and compounds.

|  | Are the elements joined together or just mixed up? | Are its properties similar to, or different from, the properties of the elements? |
| --- | --- | --- |
| Mixture of elements |  |  |
| Compound |  |  |

## Get this

- The substances in a mixture are not joined to each other.
- You can separate the substances in a mixture.
- A mixture has properties like those of the substances in it.
- The amounts of the substances in a mixture can change.

### Oxygen – a vital element

You can't live without the element oxygen. The air you breathe contains 21% oxygen mixed with other gases. Your lungs separate oxygen from these gases.

You can buy tins of oxygen. The tins contain 95% oxygen gas mixed with flavourings. People say that breathing from the tins makes them feel energetic and helps them concentrate.

**Learn about**
- Elements, compounds and mixtures that are vital for life

### Water – a vital compound

Water is a compound. It's vital to life. When it comes out of the tap it's mixed with small amounts of other substances.

Bottled water is also a mixture. The label shows that compounds containing 37 mg of sodium are dissolved in 1 litre of water.

**?** **1** Name **one** element and **one** compound that you can't live without.

**2** The water in the bottle is mixed with particles of several other substances. Name three of these substances.

**3** Suggest how you could separate pure water from the substances that are mixed with it. Use page 92 to help you.

TYPICAL ANALYSIS: Mg/l
Calcium 44.0    Magnesium   6.8
Sodium   37.0   Chloride    13.0
Nitrate  15.0   Fluoride    0.09
Sulphate 10.9   Potassium   2.0
Total Dissolved Solids: 214
Composition in accordance with the results of the recognised analysis of 23.11.01

### Other important compounds

As well as water, your body needs proteins, carbohydrates, fats and minerals. These substances are all compounds.

To be healthy, you also need small amounts of elements like iron and calcium. But it's no use swallowing iron nails or a lump of calcium metal! You need compounds that contain the elements. These compounds are called **minerals**.

| Mineral containing... | The mineral is in... | You use it to... |
|---|---|---|
| Calcium | milk, cheese, yogurt, eggs, green vegetables | make bones and teeth |
| Iron | meat, beans, some breakfast cereals | carry oxygen in the blood |
| Phosphorus | meat, fish, milk, eggs, vegetables | make bones and teeth and to help muscles work well |

**4** Name **two** metal elements and **one** non-metal element that your body needs. Give the symbols of these elements. Use page 80 to help you.

**5** What would happen if you tried to swallow a lump of the element calcium instead of a compound that contains calcium? Use page 84 to help you.

**6** Iron sulfate is a mineral. Name **two** of the elements in iron sulfate. Are these elements joined together or just mixed up?

## Studying minerals in food

Scientists have kept records of the amounts of minerals in foods for many years. Recently, a scientist studied these data. He concluded that meat and milk contain smaller amounts of minerals now than they did in 1940.

| Element | Average amount of the element in 100 g of beef (mg) | | Average amount of the element in 100 g of milk (mg) | |
|---------|------|------|------|------|
|         | 1940 | 2002 | 1940 | 2002 |
| Iron    | 6.0  | 2.7  | 0.08 | 0.03 |
| Calcium | 5.2  | 5.0  | 120  | 118  |

**7** Describe one way of finding an average.

**8** Between 1940 and 2002, did the average amount of iron in beef get larger, smaller or stay the same?

**9** What happened to the average amount of calcium in milk between 1940 and 2002?

**10** Some scientists say that the differences are because animals ate different food in 1940 and 2002. Others say that the differences are because scientists made more accurate measurements in 2002 than they did in 1940. Which reason do you think is more likely? Why?

**11** Imagine you discovered that the 1940 data was based on 10 000 cows and the 2002 data was based on 1000 cows. Which year's data do you think are more reliable? Why?

**12** The scientist who studied the data makes and sells mineral supplements. Does this change your opinion about his conclusion? What extra information would you like to help you decide whether to believe the scientist's conclusion?

## Get this

- These substances are vital to life:
- the element oxygen
- the compound water
- minerals (compounds that contain elements that the body needs in small amounts).

What links the pictures below? They show places where **chemical reactions** happen. In fact, chemical reactions are happening everywhere, all the time, even inside us!

## Learn about
- What chemical reactions are
- How to recognise them
- How they're different to reversible changes

Scientists study chemical reactions in labs, too. They use them to develop medicines, fuels and materials.

All chemical reactions:

- **create new substances** – the substances you end up with are different from the ones you started with
- are **irreversible** – at the end of the reaction, you can't get back the substances you started with.

**? 1** Decide how you can tell that each of these changes is a chemical reaction.

I reckon any type of burning is a chemical reaction.

Doesn't cooking involve chemical reactions?

Chemical reactions happen inside a plant when it makes its food from carbon dioxide and water.

When I was 7, I left a tooth in cola for a week. It got much smaller. That was a chemical reaction!

In Year 6, we mixed some vinegar with bicarbonate of soda. It made loads of bubbles. That change was irreversible.

## The signs of a chemical reaction

There are lots of clues to look out for. You might:

- see huge flames … or tiny sparks
- notice a sweet smell … or a foul stink
- feel the chemicals getting hotter … or colder
- hear a loud bang … or gentle fizzing.

It's getting hotter!

POP

And by the end of the reaction, what you see probably looks very different to what you started with.

**2** Look again at the changes in question 1. For each change, write down the signs that show it is a chemical reaction.

## But remember... not all change is chemical

Some changes are not chemical reactions – even though they might seem to be. If you heat liquid water for long enough, it boils and turns into steam. But steam is just water in a different form. When it hits a cold window it turns back to liquid water.

This is a **reversible** change. You can get back the chemicals you started with.

**3** How you can tell that each of these changes is a reversible change, not a chemical reaction?

- melting chocolate
- dissolving sugar in tea
- freezing water to make ice

## Summing up

Decide whether each of these changes is a **chemical reaction** or a **reversible change**. Give a reason for each decision.

**4** Burning gas in a cooker.

**5** Dissolving coffee powder in hot water.

**6** Boiling water to make steam.

**7** Cooking a carrot.

**Get this**

- Chemical reactions make new substances.
- Chemical reactions are irreversible.
- Reversible changes are not chemical reactions.

### Distillation

It rains very little in Saudi Arabia. There are no lakes and rivers there, either, so most people rely on water from the sea. But you can't drink seawater!

Factories like this supply water to Saudi Arabian homes. The factories take in seawater. They separate drinking water from the salts that are dissolved in it. This works because the salts and water have not been joined together in a chemical reaction. The salts dissolved in water (the solvent) in a reversible change. Seawater is a mixture.

**1** How can you get tiny grains of sand out of seawater?

**2** How can you get salt out of seawater?

**3** Why is it possible to separate drinking water from the salts dissolved in seawater?

One way of separating drinking water from salts is **distillation**. In distillation, pure water evaporates from seawater as steam. The steam cools down. It condenses into pure liquid water – without the salt.

**Learn about**

Separating mixtures made in reversible changes by:
- Distillation
- Chromatography

### How do we show... that you can get pure water by distillation?

The salty water boils. Steam evaporates from the water. Steam travels through the condenser and cools down. When it cools it condenses back to liquid water. This water goes into the beaker.

Solid salts are left behind in the round-bottomed flask.

**4** What makes the water boil?

**5** Why does steam condense to liquid water when it travels through the condenser?

**6** Where are the salts at the end of the experiment?

**7** The salts don't evaporate. Suggest a reason for this.

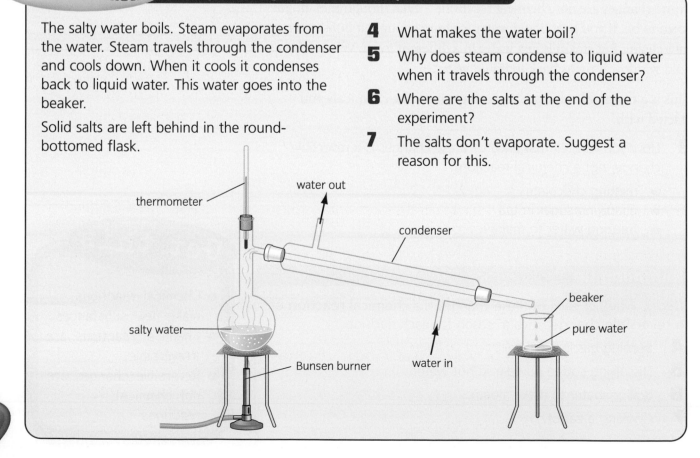

## Chromatography

You can use **chromatography** to separate mixtures, too. Chromatography works if all the substances in the mixture dissolve in one solvent.

The **chromatogram** on the right shows that the green felt tip pen contains a mixture of dyes.

**8** What colour dyes are mixed together in the green felt tip pen?

**9** Which dye travels furthest?

**10** Name the solvent in the experiment.

The blue dye goes further up the paper. This might be because the blue dye dissolves better in the solvent. Or it might be because the yellow dye sticks more strongly to the paper. You can't tell just by looking at the chromatogram.

Rashid ground up a spinach leaf in a pestle and mortar. He put a spot of spinach juice near the bottom of a piece of chromatography paper. Then he stood the paper in a solvent. He obtained the chromatogram shown above right. It shows the pigments (colours) in spinach.

**11** How many different pigments are in the spinach juice?

**12** Which pigment travelled least far up the paper?

**13** Which pigment probably dissolves best in the solvent?

chromatography paper

pencil

beaker

water

▲ Chromatogram of a green felt tip pen.

▲ Chromatogram of a spinach leaf.

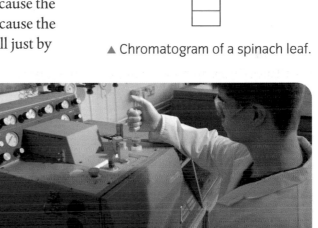

▲ Gas chromatography.

## Using chromatography

There are different types of chromatography. All are really useful.

- Police forces use gas chromatography to find out how much alcohol is in a blood sample from a driver. This shows whether the person was drink-driving.
- Detectives use chromatography to look for traces of explosives on the body hair of suspect bombers.

### Summing up

**14** Draw and label a diagram to explain how to separate pure water from a solution of salt in water.

**15** Describe **one** use of chromatography.

**Get this**

- Distillation separates solvents from solutions.
- Chromatography separates mixtures if the substances dissolve in the same solvent.

**Learn about**
- What happens when things burn
- Word equations

### Burning reactions

Have you eaten hot food this week? Used electricity? Travelled by bus, train or car? If so, you've benefited from **burning**. Burning gas to cook food is a chemical reaction. So is burning coal to make electricity – or diesel to run a bus.

Burning reactions are incredibly useful. They usually get very hot. We can use the heat to do all sorts of things.

**1** List **four** useful burning reactions.

**2** Why are burning reactions useful?

### Burning wood

Danielle is camping. She makes a wood fire to heat her baked beans. Burning wood is a chemical reaction. The wood reacts with **oxygen** from the air.

When Danielle goes to bed, she pours water on the fire to put it out. In the morning, all that's left is a pile of ash. She has to collect more wood to cook toast for breakfast.

**3** Give **two** pieces of evidence that show that burning wood is a chemical reaction.

**4** Why do you think that pouring water on the fire makes it go out?

### How do we know... that burning needs oxygen?

Air is a mixture of gases, including nitrogen, oxygen, argon and carbon dioxide. So how do we know that burning needs oxygen and not one of the other gases?

**5** How does the investigation show that oxygen is important for burning?

**6** Does the investigation prove that nitrogen is not involved in burning? How do you know?

**7** Suggest **one** way of improving the investigation.

Watch what happens to a glowing splint in air...

...and in oxygen.

## What do burning reactions make?

Burning wood makes a mixture of substances, including ash, smoke and gases we can't see. The substances made in chemical reactions are **products**.

▲ A product of a burning reaction.

Some sparklers contain tiny pieces of iron. When they burn there is a chemical reaction. Iron reacts with oxygen. Iron and oxygen are the **reactants** – the substances that react together.

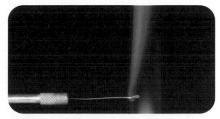

▲ A burning calcium compound.

The product of the sparkler reaction is **iron oxide**. This is the black stuff that's left when the sparkler finishes burning. The word *oxide* comes from *oxygen*. So the name *iron oxide* shows that it contains iron and oxygen. Scientists call burning **combustion**.

**8** Name the product made when iron burns.

**9** Another sparkler contains magnesium. What is the product of the chemical reaction of magnesium with oxygen?

**10** A third sparkler contains aluminium. Name the reactants of the burning reaction that happens in this sparkler.

▲ A burning potassium compound.

## Showing reactions simply

You can use **word equations** to represent reactions. They show:

- starting materials (reactants) on the left
- **products** on the right.

This is the word equation for burning magnesium:

$$\text{magnesium} + \text{oxygen} \rightarrow \text{magnesium oxide}$$

**11** Name the reactants in the word equation above.

**12** Name the product in the word equation above.

**13** Write word equations for the combustion reactions of these substances:  iron  calcium  zinc  potassium

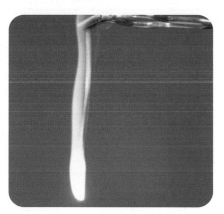

▲ Burning magnesium.

### Summing up

**14** When a substance burns, what gas does it react with?

**15** What is the scientific word for burning?

**16** Name the product made when sodium burns.

**17** Write a word equation for the burning reaction of aluminium.

**18** Name the reactants in this equation:

$$\text{sulfur} + \text{oxygen} \rightarrow \text{sulfur dioxide}$$

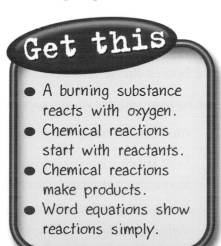

### Get this

- A burning substance reacts with oxygen.
- Chemical reactions start with reactants.
- Chemical reactions make products.
- Word equations show reactions simply.

### The wheels on the bus go round and round

You cooking chips on here? This bus stinks!

No. Our company wants to use less diesel. So this bus runs on cooking oil instead.

A few buses run on cooking oil. But most are fuelled by diesel. Diesel is a mixture of **hydrocarbons**. Hydrocarbons are compounds. They are made of atoms of two elements – hydrogen and carbon – joined together (see page 84).

A bus only goes when its fuel is burning. Energy from the burning reaction makes the engine work. Burning a hydrocarbon fuel makes new substances, including carbon dioxide and water. These come out of the exhaust pipe.

**1** Why do buses burn fuels – to make new substances or to provide energy?

**2** When a fuel burns, what gas from the air does it react with?

**3** Name **two** products that are made when diesel burns.

**4** In a bus, what happens to the products of the burning reaction?

**5** The fuel for many cars is petrol. Petrol is a mixture of hydrocarbons. Name **two** products made when petrol burns.

### Magnificent methane

Natural gas – **methane** – is a compound made of hydrogen and carbon atoms. So it's a hydrocarbon – and a very useful one! Gas cookers burn methane. Many central heating boilers burn methane to heat homes. Bunsen burners burn methane gas too.

The equation shows what happens when methane burns:

methane + oxygen → carbon dioxide + water

**6** Give **three** uses of methane.

**7** Name the reactants in methane's burning reaction.

**8** Name the products of methane's burning reaction.

**9** When gas burns in a cooker, what do you think happens to the products of the reaction?

## Greenhouse gases

You can't usually see the substances made when hydrocarbons burn. But they don't just disappear!

Carbon dioxide gas and water vapour spread out by diffusion, and mix with other gases in the atmosphere. This is not good news. Both carbon dioxide and water are **greenhouse gases**. They help to make the Earth hotter. This causes many problems (see page 122).

**10** Where do the products of burning reactions end up?

**11** Why is this a problem?

## Where do hydrocarbons come from?

Companies make diesel from crude oil. Crude oil, natural gas and coal are **fossil fuels**. Fossil fuels are the remains of animals and plants that died millions of years ago (see page 122).

Engineers drill deep wells to get crude oil and gas from under the ground or sea. Crude oil is a thick, smelly black liquid. It is mainly a mixture of hydrocarbons. Companies use distillation to separate the hydrocarbons in the mixture. They use some of the hydrocarbons to make petrol and diesel.

◄ A North Sea oil rig.

**12** What is crude oil?

**13** What process do companies use to separate mixtures of hydrocarbons?

### Summing up

**14** Where do the hydrocarbons in petrol and diesel come from?

**15** Why do cars burn petrol – to provide energy or to make carbon dioxide?

**16** Butane is a fuel for camping stoves. It is a hydrocarbon. Write a word equation for the combustion reaction of butane

*Brainache*

**Q** Are coal and cooking oil hydrocarbons?

**A** No. Coal is just carbon. The compounds in cooking oil contain oxygen as well as carbon and hydrogen.

*Get this*

- Diesel, petrol and methane are hydrocarbon fuels.
- Burning hydrocarbons makes carbon dioxide and water.

# More useful chemical reactions

## Air bags

Sam's car had airbags. Inside each airbag were three chemicals. The chemicals reacted when the car crashed and made nitrogen gas. This gas filled the airbag. So Sam didn't hit the steering wheel in the crash – just the inflated airbag.

I was in a car crash. I smelt burning and heard a loud bang. I thought the car was exploding. But it wasn't – it was just the air bags inflating.

**?** **1** How do you know this is a chemical reaction? Give **two** pieces of evidence.

**2** Name the useful product of this reaction.

## Photosynthesis

Without chemical reactions there would be no life. One vital reaction happens in plants. Plants cannot eat. So they make their own food in a chemical reaction called **photosynthesis**.

This apple tree takes in water through its roots. The water travels to the leaves. The tree takes in carbon dioxide gas from the air straight into its leaves. Here, a chemical reaction happens. The tree uses energy from light to make the water and carbon dioxide react. The reaction makes two new substances – **glucose** and oxygen gas.

Glucose is the tree's 'food'. The tree uses the glucose to stay alive, to grow and to make seeds and fruit. The tree doesn't need some of the oxygen gas. So it passes out of its leaves and back into the air.

The word equation for photosynthesis is:

water + carbon dioxide $\rightarrow$ glucose + oxygen

**?** **3** Give **one** piece of evidence that shows that photosynthesis is a chemical reaction.

**4** Name the reactants and products of photosynthesis.

Plants make their own food, so we don't have to. We eat plants. Many people also eat animals, or things that animals make, like milk or honey. These animals eat plants too.

**?** **5** Which part of the apple tree do humans eat?

**6** If you cut down the apple tree, what part of it could you use? What could you use it for?

**7** What plants – or parts of plants – have you eaten today?

## Materials for clothes

Do any of your clothes contain polyester, or nylon? These materials are made from substances separated out of crude oil (the same thick, black liquid that petrol and diesel come from!).

To make polyester, scientists start with two reactants made from crude oil: ethane-1,2-diol and benze-1,4-dicarboxylic acid. In a chemical reaction, two products are made – polyester and water. The word equation below shows the reaction simply:

ethane-1,2-diol + benzene-1,4-dicarboxylic acid → polyester + water

**8** What process do companies use to separate substances from crude oil? (see page 97)

**9** Name the products of the chemical reaction shown in the equation above.

**10** Do you think you could make ethane-1,2-diol from a polyester shirt? Give a reason for your decision.

Chemical reactions make natural fibres too – like wool, cotton, linen and silk. So chemical reactions in plants make cotton and linen, sheep make wool in a chemical reaction, and chemicals in silkworms react to make silk.

▲ A cotton plant.

### Summing up

**11** Name the two substances that react together in photosynthesis.

**12** One product of photosynthesis is glucose. Name the other product.

**13** What do plants use glucose for?

**Get this**

- Photosynthesis is a chemical reaction that happens in plants.
- In photosynthesis, carbon dioxide and water react together to make glucose and oxygen.
- Scientists make useful products in many other chemical reactions.

## 8.6 Global warming

Learn about
- What causes global warming
- What its effects are

### What's the problem?

All scientists agree that the Earth is getting hotter. This is **global warming**. By 2100 the average temperature of the Earth may be between 2 and 5 °C hotter than it is now.

Scientists from many specialisms predict what will happen when the Earth warms up.

I'm a biologist. I study polar habitats. I predict that by 2050 polar bears will be an endangered species.

I'm an agriculturalist. I predict that by 2020 the climate in some African countries may be so hot and dry that food crop harvests are halved.

I'm an oceanographer. Ice is melting in the Arctic and Antarctic. And seawater expands when it heats up. So I predict that sea levels will rise. Millions of people living near the sea may lose their homes – and lives – to flooding.

**?**

**1** Give **three** problems that global warming might cause.

**2** What do you think an oceanographer studies?

**3** Why are scientists from many specialisms interested in global warming?

**4** The scientists cannot be sure that their predictions are correct. Why?

**5** Do you think it's up to scientists to decide what to do about global warming? Why?

### Carbon dioxide in the atmosphere

Carbon dioxide gas has been part of the Earth's atmosphere for millions of years. It helps keep the Earth warm. Plants use it to make their food. Without carbon dioxide there would be no life on Earth.

Now, nearly all scientists agree that increasing amounts of carbon dioxide in the atmosphere cause global warming. But why is there more carbon dioxide in the atmosphere now than ever before? One big cause is the increase in the burning of fossil fuels like coal, oil and natural gas (methane) to heat houses and make electricity.

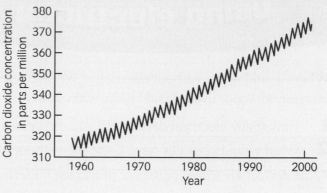

**6** Why do you think there has been an increase in the burning of fossil fuels?

**7** Explain why this adds to the amount of carbon dioxide in the atmosphere.

**8** Write a word equation for the reaction of burning methane.

**9** What does the graph show?

**10** Measurements made by many scientists were used to plot the graph. Why couldn't one scientist have collected all the data?

## Global warming – taking action

Everything you do adds carbon dioxide to the atmosphere – even breathing! The table shows how much carbon dioxide is produced by travelling.

| For one person to travel one mile: | Approximate amount of carbon dioxide made in g |
|---|---|
| in a 'family diesel' car | 250 |
| by train | 40 |
| by bus | 170 |
| by aeroplane | 600 |

**11** Why are the amounts of carbon dioxide only approximate?

**12** Which method of transport makes the most carbon dioxide?

**13** Which method of transport in the table makes least carbon dioxide? Can you think of a method of transport that makes even less carbon dioxide than this?

Generating electricity also makes lots of carbon dioxide. You can find out about this in Unit 12 *Energy*.

The amount of carbon dioxide a person's activity produces in a year is their **carbon footprint**.

**14** The average carbon footprint of a British person in 2007 was 11 000 kg. The average carbon footprint in 1907 was much less than this. Suggest **two** reasons for the difference.

**15** Suggest **three** ways of reducing your carbon footprint.

*Brainache*

**Q** Is carbon dioxide the only greenhouse gas?

**A** No, methane is a greenhouse gas, too. One cow burps 280 kg of methane into the air every year!

*Get this*

● Extra carbon dioxide in the atmosphere causes global warming.
● Global warming may cause flooding, extinctions and poor harvests.

Where would we be without electricity? We take it for granted but imagine what our lives would be like without it.

**1** How many electrical devices did you use yesterday?

**2** What would your day have been like without electricity?

Some devices are mains powered; others use batteries.

The devices shown here are all battery powered.

**3** Give **one** advantage of battery-powered equipment.

**4** Give **one** disadvantage of battery-powered equipment.

**Learn about**
- How circuits work
- How to draw circuit diagrams
- Series circuits

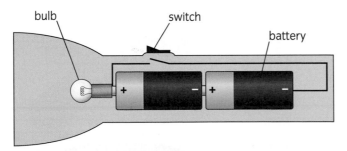

## Electric circuits

For a device to work we need a complete path, or **circuit**. The diagram shows the inside of a torch.

Here is an electric **cell**. A **battery** contains two or more cells. It gives more energy than a single cell.

bulb    switch    battery

**5** Look at the pictures of electric circuits below. In which circuit, A, B or C, will the bulb light when the switch is closed?

**6** What must you do to make the other bulbs light?

A    B    C

## Circuit diagrams

We draw a circuit diagram to show how the **components** are connected. This uses standard **symbols** instead of drawing a picture.

| Component | Symbol |
|---|---|
| cell | 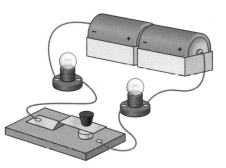 |
| battery | |
| bulb | |
| switch | |
| connecting wire | |

 **7** Why is it better to use circuit symbols than drawing the circuit just the way it looks?

This is a **circuit diagram** for the torch. The switch in the circuit diagram is in the 'open' position.

**8** What must you do to make the bulb light?

**9** Draw a circuit diagram for the arrangement shown.

**10** Sam adds another bulb. What does she notice?

**11** Sam then adds another cell. What happens?

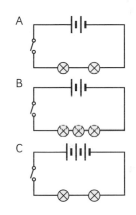

## Series circuits

We connect the components in the torch in **series**. They are joined together in a loop. If a circuit contains one cell and one bulb the bulb is lit to **normal brightness**.

The torch contains *two* cells joined in series.

**12** Describe the brightness of the torch bulb.

**13** In which circuit, A, B or C, will the bulbs be brightest?

**14** In which circuit, A, B or C, will the bulbs be dimmest?

The bulbs cannot be switched on and off separately. Either all light or none of them lights. If one bulb 'blows', or is taken out of its holder, all the bulbs go out.

 **15** Why do you think this happens?

### Summing up

**16** Complete the sentences:
- If you add *more bulbs* to a series circuit the bulbs become ...
- If you add *more cells* to a series circuit the bulbs become ...

**17** Draw circuit diagrams for the three pictures in question 5.

**18** Draw a diagram of a series circuit that will allow three bulbs to light to normal brightness.

**Get this**

- Circuits must be complete to work.
- We show circuits using a circuit diagram, with standard symbols.
- In a series circuit components are joined in a single loop.

### Parallel circuits

A **parallel** circuit is a much more useful type of circuit than a series circuit.

Each bulb in a parallel circuit forms a complete circuit with the cells. This means two bulbs light to the same brightness as one. If you add more bulbs the brightness of all of them stays the same.

In this diagram, when bulb Y is added to the circuit bulb X stays as bright as it was before, and bulb Y lights to the same brightness as X.

 **1**   What happens when we add bulb Z to the circuit?

If one bulb 'blows', or is taken out of its holder, the other bulbs still light. You can switch the bulbs on and off independently.

**2**   In this circuit, which bulb or bulbs will light up if you close switches A and C?

**3**   Which bulb or bulbs will light up if you close switches B and C?

**4**   Do you think the lights in your house are wired in series or in parallel? Explain your choice.

### Learn about

- Series and parallel circuits
- Electric current
- How to measure electric current

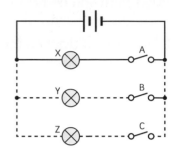

▲ Adding a third bulb, Z.

Lights out everybody!

### The inside story

Circuits work because an **electric current** goes through them. We measure electric current in **amperes** (**amps** or **A** for short) using an **ammeter**.

 **5**   Draw a circuit diagram for a bulb and cell.

**6**   This is the circuit symbol for an ammeter. –(A)–
Add it into your circuit for question 5 to measure the current.

Every time you add another bulb the ammeter reading decreases.

 **7**   What would you notice about the brightness of the bulbs?

**8**   If you added a third bulb to this circuit what would happen to the current?

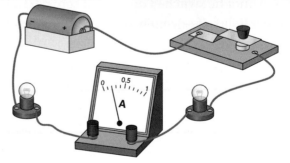

▲ An ammeter wired into a series circuit.

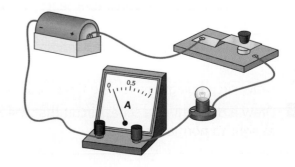

In a series circuit the current is the same everywhere in the circuit. The readings are the same on the ammeters at P, Q and R.

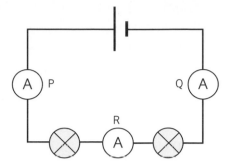

 **9**  If the ammeter at R shows a reading of 0.1 A, what will the ammeter at P show?

## Current in parallel circuits

The current in a parallel circuit splits up through all the branches. In the diagram it splits as it reaches X. Some goes through bulb $L_1$ and the rest through bulb $L_2$. The two currents join up again at Y and travel back to the cell. So:

**Current at L = current at M + current at N**

You can think of this like cars on roads. Cars may choose to take different routes but the total number stays the same.

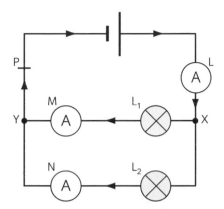

If the bulbs are the same power, the current at M and N will be the same.

 **10**  If the current at each of M and N was 0.2 A what would it be at L?

**11**  If the current at N was 0.1 A and at M was 0.2 A what would it be at L?

**12**  If there was an open switch at P which bulb or bulbs, if any, would be lit?

**Q** What is electric current?
**A** Current is a flow of electric charge (electrons) through a conductor like metal.

## Summing up

**13**  Copy and complete the table summarising series and parallel circuits. The first part has been done for you.

| Series | Parallel |
|---|---|
| Bulbs get dimmer when you add more bulbs. | Bulbs stay at the same brightness when you add more bulbs. |
| The current in each bulb gets less when you add more bulbs. | |
| | You can switch the bulbs independently. |
| If one bulb 'blows' they all go out. | |

**14**  You are designing a circuit to fit spotlights in your bedroom. Which type of circuit would you use?

## Get this

- We measure electric current in amperes (A) using an ammeter.
- The current is the **same** all around a series circuit.
- In a parallel circuit each bulb forms a complete circuit with the cells.

**Learn about**
- What 'voltage' means
- The difference between current and voltage

### Keeping the current moving

Energy is needed to push the current around a circuit. A cell provides this energy. So does the electric mains. This energy is called **electrical energy**.

Electrical energy also makes the components in a circuit work. Some of the electrical energy becomes heat energy in a kettle or sound energy in an MP3 player.

**?** **1** What form of energy does electrical energy become in a battery-operated car?

### Measuring energy

**Voltage** tells us about the energy provided by a battery or the mains. Voltage is measured in **volts** using a **voltmeter**.

 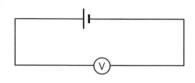

The picture and diagram show how you would connect your voltmeter to show how much energy a cell puts into a circuit. This is called the cell's voltage.

The higher the voltage the more energy the components can use.

**?** **2** When you buy a battery, how can you tell which one will provide most energy?

A voltmeter also shows how much electrical energy is changed into other forms such as light and heat by a component like a bulb.

It is connected in the same way, in parallel across the bulb. The voltage across the bulb is 6 volts.

If you added another bulb of the same sort to this series circuit at P and measured the voltage across both, each bulb would show a voltage of 3 V. The energy from the battery is shared between them.

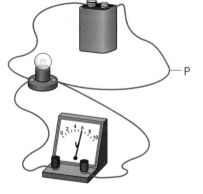

**?** **3** How does this explain why bulbs in a series circuit get dimmer every time you add another one?

**4** If you added a third bulb to this 6 volt circuit, what do you think the readings would be on voltmeters across each one?

### Parallel magic

In a parallel circuit each bulb makes its own circuit with the cell or battery. So the voltage is not affected by the number of bulbs in the circuit.

**5** If you added another bulb to this circuit what would the voltage across it be?

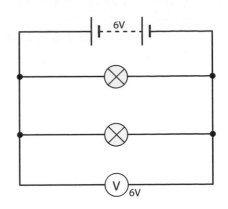

## Current and voltage

Why does voltage change across components but not current?

Current is a flow of **electrons** through the metal wires. Electrons are tiny particles in the metal that move around randomly. A battery gives the electrons energy and they flow through the wire in the same direction. This is electric current. When the battery is switched off the electrons move randomly again.

Electrons in a wire *without* a battery connected ...

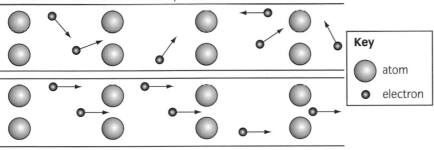

**Key**

⬤ atom

∘ electron

... and *with* a battery connected.

A voltmeter compares the amount of energy the electrons have before and after passing through the bulb. When the electrons get back to the battery they are given more energy to go round again.

 **6** When does current stop flowing in a circuit?

**7** Current in a circuit is never used up. Why?

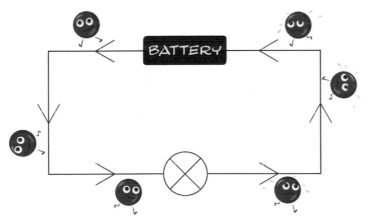

### Brainache

**Q** Does the voltage of a battery affect the amount of current?

**A** Yes, the higher the voltage the higher the current.

## Summing up

**8** How many 1.5 V cells would you need to provide a voltage of 9 V?

**9** Dilip connected two bulbs in parallel with a 9 V battery. Draw his circuit diagram. Add a voltmeter to measure the voltage across one of the bulbs. What would it read?

### Get this

● Electrical energy:
— pushes the current around a circuit
— is changed to other forms in components.

● Energy is used up in a circuit but current is not.

### Learn about
- Properties of magnets
- Magnetic materials
- Magnetic fields

### What are magnets?

Fridge magnets, like these smiley faces, stick to the steel door of your fridge.

A long time ago the Ancient Greeks discovered a rock which they called **lodestone**. Pieces of this rock sometimes **attracted** each other. Sometimes they pushed each other away, or **repelled**.

A magnet settles in a North–South direction when hung by a thread. It has a **pole** near each end – North-seeking and South-seeking.

**1** How does this explain the behaviour of the lodestone found by the Ancient Greeks?

▲ Lodestone is a natural **magnet**.

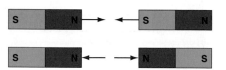

▲ Like poles repel, unlike poles attract.

### How do magnets work?

**Magnets** attract **magnetic** materials. **Iron**, **steel**, **nickel** and **cobalt** are magnetic materials. Magnetic materials are not attracted to each other but they can be made into magnets.

- **Iron** is *easy* to magnetise but *loses* its magnetism easily.
- **Steel** is *hard* to magnetise but *keeps* its magnetism.

Jack dips a magnet into a pot of iron filings. Each tiny iron filing becomes **magnetised**. It's now a tiny magnet with a North and a South pole. This means it can attract more iron filings that then become magnetised.

**2** How does the photo show that the poles are near the ends of a magnet?

### How do we know... that some materials keep their magnetism?

Try picking up a line of iron nails with a magnet. Then carefully take the magnet away.

Repeat this using steel pins instead of iron nails.

**3** What material, iron or steel, do you think bar magnets can be made from?

**4** What are the poles marked X and Y in the diagram?

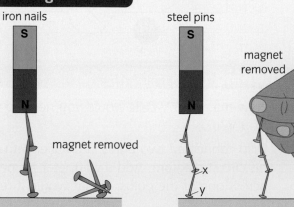

iron nails    steel pins

magnet removed

## Magnetic fields

The effect of a magnet is felt in the space around it. A magnet attracts a pin without touching it. Try it! We call this region around a magnet its **magnetic field**.

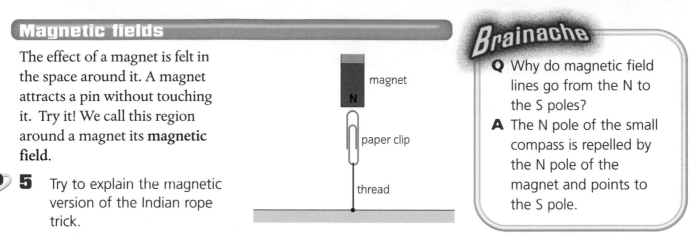

**5** Try to explain the magnetic version of the Indian rope trick.

You can use small **compasses** to show the magnetic field.

Where the lines are closer the magnetic field is stronger. Magnetic field lines go from the N pole to the S pole.

You can get a general idea of the shape of the magnetic field using iron filings. At X the magnetic fields of each magnet cancel out. X is a **neutral point**.

**6** What can you say about the poles of the magnets giving the magnetic field shown?

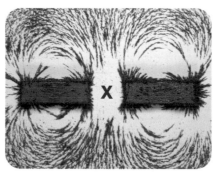

### Brainache

**Q** Why do magnetic field lines go from the N to the S poles?

**A** The N pole of the small compass is repelled by the N pole of the magnet and points to the S pole.

small compasses

North pole

South pole

## Summing up

**7** Which of these materials are magnetic?

brass   cotton   iron   nickel   wood   steel

**8** Copy the diagrams and draw the shape of the magnetic field around the bar magnets shown. Mark the position of any neutral points with the letter X.

A

S   N      S   N

B

N   S      S   N

**9** Arani hid a small bar magnet in a matchbox. Use the plotting compass directions shown to decide the position of the magnet.

Laddie's own
Safety
Matches

### Get this

- Like poles repel, unlike poles attract.
- A magnetic field is the region around a magnet.

### What are electromagnets?

In **electromagnets** the magnetism can be switched on and off by switching an electric current on and off. This makes them very useful. For example electromagnets attached to cranes are used to move wrecked cars around in a scrap yard.

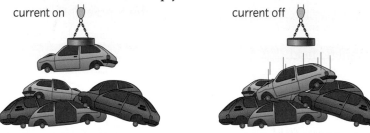

current on　　　　　　current off

### How are electricity and magnetism linked?

When an electric current passes through a long straight wire it creates a **magnetic field** around it.

 **1**　What change do you think there would be in the magnetic field if the current went *down* instead of up the wire?

When an electric current passes through a coil of wire, called a **solenoid**, it creates a magnetic field around it. The solenoid behaves like a magnet.

 **2**　What does the shape of the magnetic field around the solenoid remind you of?

If an iron rod is put inside the solenoid the iron rod becomes magnetised. The iron rod is called an iron **core**. The core makes the magnetic field stronger. The solenoid and core together make an electromagnet.

 **3**　Why is iron used as the core of an electromagnet?

**4**　What would you expect to happen if you put a steel rod inside the solenoid instead of an iron one?

### Making a stronger electromagnet

A stronger electromagnet has a greater magnetic attraction so can hold a bigger load. You can increase the strength of an electromagnet by:

- increasing the size of the current
- winding more turns on the solenoid.

There is a limit to the strength of an electromagnet.

**5**　Suggest how you could test the strength of an electromagnet.

### Learn about
- The magnetic effect of an electric current
- How to make electromagnets
- How to change the strength of an electromagnet
- Uses of electromagnets

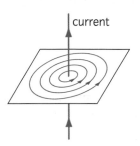

current

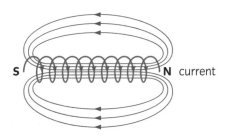

S　　　　　　　N　current

#### Brainache

**Q** Why does an iron core make the magnetic field stronger?

**A** The magnetic field of the magnetised iron core is added to that of the solenoid.

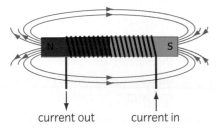

current out　　current in

## Using electromagnets

Some door bells contain an electromagnet.

When the bell switch is pushed the circuit is completed.

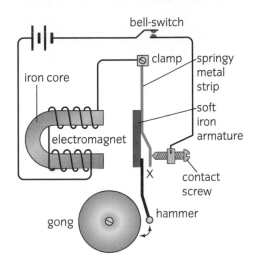

- A current passes through the solenoid and the iron core is magnetised.
- The electromagnet attracts the soft iron armature and the hammer hits the gong.
- This breaks the circuit at X so the electromagnet is no longer magnetised.
- The armature springs back, completing the circuit.
- The sequence starts again so the bell keeps ringing.

**6** Suggest why this is often called a 'make and break' circuit.

Some trains move along without wheels thanks to magnetism. Powerful electromagnets have been used to develop high-speed trains, called **maglev trains**.

Maglev is short for magnetic levitation. Maglev trains can travel at speeds of up to 310 mph (500 kph). Metal coils line a guideway or track, making electromagnets. There are large guidance magnets attached to the underside of the train.

The train floats over the guideway on a 'cushion' of air which eliminates friction between the train and the track.

**7** Why can Maglev trains go faster than other trains?

Sometimes doctors use electromagnets to remove iron or steel splinters from eyes.

**8** Why is this only suitable for removing iron or steel splinters?

**9** How many other things can you think of that use an electromagnet?

### Summing up

**10** Give **two** ways of increasing the strength of an electromagnet.

**11** Explain why an iron core is better than a steel core for an electromagnet.

**12** Describe how an electromagnet is used to move cars around in a scrap yard.

**13** Why is an electromagnet, rather than a permanent magnet used to remove a steel splinter from a metal-worker's eye?

**Get this**

- In electromagnets the magnetism can be switched on and off.
- An electric current in a wire creates a magnetic field.
- You can make a stronger electromagnet by a bigger current or more turns on the solenoid.

-Imagine living without electricity! The world would be a very different place. But all our knowledge of electricity has developed in the last 150 years or so. Find out what life was like 200 years ago. Ask your parents and grandparents what electrical equipment they had when they were your age.

**Learn about**
- Creative thinking in science
- Using evidence to test ideas
- Developing theories

**?1** Which electrical device would you find it hardest to do without?

## All about frogs

During the 1790s, Italian doctor **Luigi Galvani** made frog muscles twitch when he touched them with two different metals joined together in an arc. He found that electric charges could make frog legs jump even if the legs were no longer attached to a frog.

Galvani was convinced that he was seeing the effects of what he called animal electricity – the life force within the muscles of the frog. He thought it was caused by an electrical fluid that is carried to the muscles by the nerves.

One story claims that Galvani also experimented on himself, using two different metals to make an arc. When he put one end of the arc in his mouth, and the other end in the corner of his eye he saw a bright spark.

**?2** What was Galvani's theory about why the dead frog legs jumped?

**3** If the story about the experiment on himself is true, why would it have made him more certain about his theory?

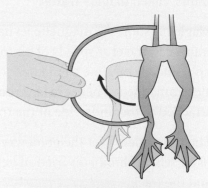

▲ Luigi Galvani

**Alessandro Volta**, Galvani's friend, did the same experiment and got the same results. This meant that Galvani's results were **reliable**, because when someone else did the same experiment they got the same results.

But Volta did *not* agree with Galvani's explanation. Volta believed the fact that the two metals were different made a current flow between them. Electric current flows through water. As the frog's flesh was full of water the current flowed through it between the metals.

▲ Allesandro Volta

**4** Why was the experiment reliable even though Volta and Galvani drew different conclusions from it?

Volta thought you could use this to bring the dead back to life. He experimented on executed prisoners. This was the inspiration for Mary Shelley's novel 'Frankenstein'.

Volta and Galvani were no longer friends! In 1794 Volta built the very first **battery** in order to disprove Galvani's theory. Volta made an electric current using metals alone with no living tissue. Volta's explanation, or **hypothesis**, was then accepted. Galvani was still cross about Volta's experiments when he died in 1798.

**5** What was Volta's explanation about why current flowed between the two metals?

Based on this theory, Volta made the first battery to produce a large current in 1799. He used bowls of salt solution connected by strips, or arcs, of copper and zinc. He had made a **voltaic pile**, the forerunner of today's alkaline batteries.

Later, Volta made a much smaller battery by building a pile of small round zinc and copper discs. He separated them from each other with paper discs that he had soaked in salt solution.

Today's batteries are much smaller and more portable!

**6** What electrical unit is named after Alessandro Volta?

**7** What other electrical unit is named after a famous French scientist?

Both Galvani and Volta were right to some extent. We now know that the signals which the brain sends through an animal's nerves to control muscles are electrical ones. So the current created between the metals was acting on the nerves in the frog's legs and making them twitch.

**8** Galvani and Volta exchanged ideas frequently. This did not often happen in the 18th and 19th centuries. Suggest why.

**9** Galvani and Volta interpreted the same evidence differently. Suggest what could be done to decide which theory is correct.

▲ Voltaic pile.

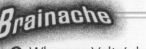

salt solution

zinc

copper

one element

▲ Volta's battery.

**Brainache**

**Q** Why was Volta's battery called a voltaic 'pile'?

**A** The French word for 'battery' is 'pile'.

**Get this**

- Scientists come up with new ideas based on what they already know or observe.
- They look for evidence to test their ideas.
- They often carry out experiments to provide evidence.

## What is energy?

Energy is needed to make things happen.

- A kettle needs energy to heat water.
- A dog needs energy to run.
- You need energy to ride your bike.
- Light bulbs need energy to light up.

I've no energy!

**Learn about**
- Some forms of energy
- Stored energy
- Energy changes

## Energy in action

Energy takes many different forms. The pictures show some of these.

The **sound** made by this trumpet is a form of **energy**.

This bike has **movement energy**. This is also called **kinetic energy**.

**Light** is a form of energy.

This bonfire is giving out a lot of energy in the form of **heat**. This is also called **thermal energy**.

The kettle uses **electrical energy** when it heats water.

**1** Write down the main form of energy which the examples listed below show.

a galloping horse
people screaming on a roller coaster
marbles rolling across the floor
a saucepan of boiling water

**2** Look at the picture of the candle. Its flame is giving out light energy but another form of energy too. What is the other form of energy?

**3** Look at the pictures of the bonfire and bike. What other forms of energy are they giving out?

## Energy stores

Where does the energy to do things come from?

We get the energy we need to do things from the food that we eat.

A car engine gets the energy it needs to move from its fuel (usually petrol or diesel).

**Coal**, **oil** and **gas** are called **fossil fuels**.

Food and fossil fuels are energy stores. They have **stored** energy in the chemicals they contain, so we call this **chemical energy**.

The energy stored up in our food, and in fossil fuels, is ready to be used when we need it. We use the energy released to make things happen. For example, the energy stored in petrol is used to make a car move.

**?** **4** Name **one** thing that the stored chemical energy in a bowl of pasta is used for.

## Energy changes

Energy changes from one form to another. For example, the runner's energy source is the food he has eaten – chemical energy. He changes some of this energy into kinetic energy when he runs.

**?** **5** Name **one** other type of energy that the runner changes his chemical energy into when he runs.

**6** What happens to the chemical energy from gas when it is ignited in a gas ring on a hob?

### Summing up

**7** Name **six** types of energy.

**8** What sort of energy do the following have?

an aircraft in flight    a slice of bread

**9** Name **three** fossil fuels.

**10** Copy and complete this sentence:

A match changes _____ energy into _____ energy and _____ energy.

**Get this**

- Some forms of energy are heat (thermal), light, sound, electrical and movement (kinetic).
- Chemical energy is stored ready for use in food and fossil fuels.
- Energy can be changed from one form to another.

## Living things and energy

All living things, plants and animals, need energy. That includes us!

We need energy for everything we do. This includes breathing and sleeping as well as running and cycling. Our hearts need energy to keep beating. Children need energy to grow bigger bones, muscles and brains.

 **1**   Name **two** other things we do that require energy.

A scientist worked out that his leg muscles only needed the amount of energy contained in a ham sandwich to climb a mountain. So that was all the food he took with him.

 **2**   Why was it a silly thing to do?

## Measuring energy

We measure energy in **joules (J)**. If you lift an apple 1 m up in the air you use about 1 J of energy. A joule is a very small amount of energy, so energy is often given in **kilojoules (kJ)**.      1 kJ  =  1000 J

**3**   You need about 2000 J of energy to walk up the stairs at home. How many kJ is 2000 J?

## Food energy

If you look at a food label you will see the energy provided by eating that food. It is in kJ as well as in an old-fashioned unit called the kilocalorie (kcal). People still talk about 'counting calories'. They should be 'counting joules'!

**4**   Which gives you more energy if you eat 100 g of it – bread or cheese ?

**5**   Why do the food labels give other information too?

## Getting the energy balance right

Ideally the energy we get from the food we eat should equal the energy we use each day.

The weight of adults should stay constant. The weight of young people who are still growing should increase, but not too much!

**Learn about**
- How living things need energy
- How we measure energy
- Getting the energy balance right
- Food and exercise

| NUTRITION INFORMATION | | |
|---|---|---|
| TYPICAL VALUES | PER 100g OF PRODUCT | PER AVERAGE SLICE (45.8g) |
| ENERGY | 1215 kJ | 556 kJ |
|  | 288 kcal | 132 kcal |
| PROTEIN | 12.3g | 5.6g |
| CARBOHYDRATE | 39.7g | 18.2g |
| of which sugars | 3.2g | 1.5g |
| FAT | 8.9g | 4.1g |
| of which saturates | 1.9g | 0.9g |
| mono-unsaturates | 3.3g | 1.5g |
| polyunsaturates | 2.6g | 1.2g |
| FIBRE | 6.0g | 2.7g |
| SODIUM | 0.49g | 0.22g |
| SALT | 1.23g | 0.55g |

▲ Bread nutrition label

USE BY: 27.08.07
KEEP REFRIGERATED BELOW +4°C

NET WEIGHT: 0.200 ℮

Batch AL500

Nutritional Information
Typical value per 100g
Energy: 410 kcal/1700kj
Protein: 25.00g
Carbohydrates: 0.1g
Fat: 3g

5 407898 001495

▲ Cheese nutrition label

**6** What other foods from the table could you eat to give you the same amount of energy as one chocolate bar?

**7** Suggest why climbers carry chocolate bars.

**8** Why does someone's weight increase when they are growing?

**9** What else should you think about when planning your meals to stay fit and healthy?

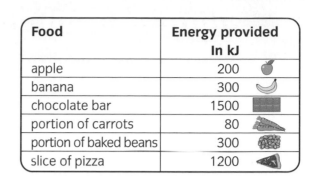

| Food | Energy provided In kJ | |
|------|------------------------|---|
| apple | 200 | |
| banana | 300 | |
| chocolate bar | 1500 | |
| portion of carrots | 80 | |
| portion of baked beans | 300 | |
| slice of pizza | 1200 | |

Some tasks require more energy than others.

▲ Sleeping – 300 kJ/hour

▲ Walking – 1000 kJ/hour

▲ Jogging – 3000 kJ/hour

**10** Which of these three people needs to eat most food to get their energy balance right?

## Getting the energy balance wrong

If we take in more joules of energy than we use our bodies store it as fat for future use. If we don't get as much energy from food or fat as we need our body may stop working properly.

Dieting means eating fewer joules than you need so your body is forced to use energy from your fat stores. So you lose weight.

We should aim to exercise and eat sensibly to keep fit and well.

**11** Slimming does not necessarily mean eating less. What is a healthier way to lose fat?

## Summing up

**12** How many joules is 15 000 kJ?

**13** Use the table above to work out how much energy you would gain if you ate a slice of pizza and a banana for lunch.

**14** For how long could you jog on this amount of energy?

### Brainache

**Q** Darren was a professional footballer. He now works in an office. Why should he change his diet?

**A** He is not as active so needs less energy from his food each day.

### Get this

- All living things need energy.
- We measure energy in joules or kilojoules.
- The energy we get from food should equal the energy we use.

**Learn about**
- Food chains
- Fossil fuels
- Energy transfers

## Food chains

Some of us are vegetarians but most of us have a **mixed diet** – we eat fruit and vegetables (plants) and meat (animals).

Plants get their energy to make food directly from the Sun. They use their food to survive but also to grow and form new plants. The plant itself is an energy store and when we eat it this energy passes to us. We can show this in a food chain.

Animals eat plants for food. They use the energy in plants for surviving and moving but also growing and forming their own bodies. This means that all parts of an animal's body store energy which passes to us when we eat it.

This means **all** our food energy comes from the Sun.

**1** How do plants use the Sun's energy to make food?

**2** Jo is eating cheese. Draw a food chain diagram for this.

**3** What do the arrows in a food chain diagram mean?

## Fossil fuels

Fossil fuels also got their stored energy from the Sun. Coal is formed from trees which grew in swamps millions of years ago. Oil and natural gas are formed in a similar way from the remains of tiny plants and animals in the oceans.

**4** What is the main difference between the formation of coal and oil?

**5** Draw an energy chain which starts with the Sun and ends with coal.

## Energy transfers

When energy is moved from one form to another we say it is **transferred**.

A table lamp **transfers** electrical energy to light energy. We can show this on an energy transfer diagram:

electrical energy → | lamp | → light energy

This New Orleans electric streetcar transfers electrical energy to kinetic energy:

electrical energy → | streetcar | → kinetic energy

**6** Draw an energy transfer diagram for an electric kettle.

**7** Draw an energy transfer diagram for a guitar.

### How do we picture... energy transfer?

Energy is **transferred** from one form to another. It may look different but it is still there, although in a different form. Energy itself is invisible – we only know it's there by what it does.

A possible model for energy transfer is football transfers. Footballers who transfer to new clubs change strip so they look different, but they are still the same underneath with the same kicking and passing skills.

**8** How is energy transfer like a football transfer?

**9** Think of **two** ways in which energy transfer is not like football transfers.

## Conserving energy

A light bulb transfers energy as light but it also heats up so it also transfers some energy as heat. This is wasted energy because we can't use it. But it is not lost. It just warms the air slightly. Energy is often wasted like this but it never disappears. We call this energy **conservation**.

### Summing up

**10** What are fossil fuels made from?

**11** Copy and complete the sentence:

A radio transfers _____ energy to _____ energy.

**12** Draw an energy transfer diagram for a hairdryer.

**13** What energy is wasted when you use a hairdryer? Explain why.

### Get this

- All the energy in plants, animals and fossil fuels comes from the Sun.
- Energy transfers from one form to another.
- Energy can be wasted but never lost.

# 10.4 Storing energy

## Potential

Sam's school report says that he has the **potential** to do very well.

Have you ever had this on your report? What does it mean?

It means Sam has the ability to do very well but he needs to work hard to make it happen.

Sometimes energy is *potential* energy.

Sam eats a sandwich for lunch. It gives him chemical *potential* energy. His body uses some straightaway but stores the rest ready for him to use later when he wants to play football.

Batteries also have chemical potential energy which can be turned into electrical energy in a circuit.

 **1** Why is it called 'chemical potential' energy?

## Gravitational potential energy

Natalie holds a ball in the air. When Natalie releases the ball it will fall to the ground because **gravity** acts on it. The falling ball has kinetic energy. This energy comes from lifting the ball up against gravity.

So when Natalie holds the ball up it has potential energy from gravity. We call this **gravitational potential energy**. Anything that is high up has gravitational potential energy (GPE for short).

When the ball falls, GPE changes to kinetic energy (KE).

 **2** Why is it called 'potential' energy?
**3** Suggest why kinetic energy is not potential energy.

Have you ever been on a rollercoaster? The carriages are pulled up to a great height by an electric motor and then released.

 **4** What sort of energy do the carriages have at their highest point?

**5** What sort of energy do the carriages gain as they fall?

The rollercoaster changes GPE to KE and back to GPE as it moves from one peak to the next.

 **6** Name **one** way in which energy is wasted in the rollercoaster.

GPE

KE

## Elastic potential energy

Stretch an elastic band. It has **elastic potential energy** (EPE).

 **7** Why is it called 'potential' energy?

**8** What happens to the elastic potential energy when you let the band go?

A stretched spring also has elastic potential energy. The **weight** is the **force** that pulls the spring down and gives it energy. When the force is removed the elastic potential energy in the spring pulls it back into shape.

A trampoline has springs around its edges to make it bouncy.

 **9** What force stretches the springs on the trampoline?

This bungee jumper has a strong elastic cord attached to his ankle. He's jumping off a bridge over a ravine.

 **10** What sort of energy does he have as he stands on the bridge?

**11** What sort of energy does he gain as he falls?

**12** The elastic cord gradually stretches. He stops at the bottom of his fall.

What sort of energy does the elastic cord have now?

The bungee jumper changes GPE to KE to EPE as he falls.

 **13** What energy changes happen as he moves up again?

He bounces up and down several times.

**14** Why does he eventually stop moving?

### Summing up

**15** Name **three** types of potential energy.

**16** Copy and complete the sentence:

A catapult transfers _____ energy to _____ energy.

**17** Maya is on the 10th floor and Trisha on the 18th floor of an apartment block. Who has the greater GPE?

### Get this

- Chemical energy can be potential (or stored) energy.
- Gravitational potential energy is energy stored by objects lifted off the ground.
- Stretched objects have elastic potential energy.

**Learn about**
- Non-renewable and renewable sources of energy

### What are non-renewable sources of energy?

Fossil fuels are **non-renewable** sources of energy.

They will run out one day fairly soon. They took millions of years to form so are difficult to replace.

Most **power stations** burn fossil fuels to make electricity.

The heat from burning is used to heat water and make steam. The moving steam turns a turbine. The turbine turns a generator which makes electricity.

 **1** What sort of energy does a fossil fuel possess?

**2** Copy and complete the energy transfer diagram to show the main energy changes that take place when a fossil fuel is burned:

$$\underline{\quad\quad} \rightarrow \boxed{\text{fossil fuel}} \rightarrow \underline{\quad\quad} + \underline{\quad\quad}$$
energy                                    energy        energy

### What are renewable sources of energy?

Most **renewable** sources depend on the Sun so they won't run out, at least as long as the Sun keeps shining. The Sun will keep on shining for another 5 billion years or so.

 **3** What's the difference between renewable and non-renewable sources?

#### Solar energy
**Solar cells** turn the Sun's energy into electricity. They are used in calculators, on illuminated road signs and on some buildings. **Solar panels** on roofs use the Sun's energy to heat water directly.

**4** What is one disadvantage of solar energy, particularly in the UK?

Solar cells can only generate small amounts of electricity. You need to turn a generator to produce large amounts of electricity.

#### Biomass energy
Biomass is the energy stored in plants from the Sun. It is used in lots of ways. Sugar from sugarcane is converted into alcohol and used as fuel for cars. A power station in Devon is fuelled by cow dung.

**5** Suggest how cow dung can be used to make electricity.

### Geothermal energy

The heat energy in hot rocks deep underground is used to heat cold water to make steam. The steam is used to make electricity as well as to provide hot water for homes.

### Wind energy

The kinetic energy of the wind is used to turn turbines which turn generators and produce electricity. Wind turbines are becoming more common but they take up a lot of room. Many people think they look ugly and are noisy.

**6** Draw an energy transfer diagram to show the main energy change in a wind turbine.

## The power of water

### Wave energy

Waves contain huge amounts of energy. Surfers use it for fun, but the up-and-down movement of the seas can be used to make electricity.

### Hydroelectricity

The gravitational potential energy of water in hilly regions is used to make electricity when the water flows downhill. A dam is often built to store water so that it can be released when needed.

**7** Copy and complete the energy transfer diagram to show the main energy changes in a hydroelectric power station.

GPE → | falling water | → _____ | generator | → _____

### Tidal energy

The height of the tides is used to turn turbines to make electricity. The river Severn has a very high tide.

## Summing up

**8** Give **three** examples of non-renewable energy sources.

**9** Give **three** examples of renewable energy sources.

**10** Suggest a suitable type of site for wind turbines.

**11** Name **one** energy resource that is not dependent on the Sun.

### Brainache

**Q** What has wind energy got to do with the Sun?

**A** Winds are caused by movement of air due to temperature changes because the Sun heats the Earth unevenly.

### Get this

- Fossil fuels are non-renewable energy sources.
- Renewable energy sources:
  - won't run out
  - mostly depend on the Sun.

### What is the energy crisis?

We are told that the world is running out of energy. But we can't lose energy. What people mean is that we are running out of *useful* energy sources such as fossil fuels.

**? 1** What is happening to the energy that we appear to be losing?

Fossil fuels contribute to **global warming** because they produce a lot of carbon dioxide. Global warming will cause all sorts of serious problems (see page 100).

### Ways of saving energy

One way of tackling this problem is to use less energy.

It's amazing how small things can save energy. For example, if everyone in the UK switched appliances off instead of leaving them on standby the energy saved in a year would be equal to the electricity produced by one whole power station.

This table shows the energy used by some other things you use all the time:

| Appliance | Energy used per hour in kJ |
| --- | --- |
| ordinary light bulb | 216 |
| energy-saving bulb | 43.2 |
| plasma TV | 990 |
| ordinary TV | 350 |
| i-pod | 100 |
| indicator light | 7.2 |

**? 2** How much energy do an ordinary light bulb and an energy-saving one use per hour in kilojoules?

**3** How much energy do you save in an hour by using an energy-saving light bulb rather than an ordinary one?

**4** How can an energy-saving bulb produce the same amount of light as an ordinary one but use much less electrical energy?

**5** Roughly how many times more energy does a plasma TV use compared to an ordinary one?

**6** Explain the message in the poster on the right.

**Brainache**

**Q** What's the ozone layer got to do with global warming?

**A** Nothing at all! It's made from a type of oxygen and protects Earth from the Sun's harmful ultraviolet rays.

IF YOU DON'T PRESERVE NATURE BY USING LOW-WATTAGE BULBS, WHO WILL?

## Why do we have an energy crisis *now*?

The Zhang family live in Beijing. They have good jobs, a nice house and want to have the things, like electrical appliances, that we already enjoy.

Until recently, most of China's people were very poor. But China is developing its manufacturing industry and this is making people richer. More electricity is needed to run factories and the electrical goods people want. In 2007 two new power stations were opened every week!

 **7** Suggest **three** things the Zhang family would like to have.

The Zhangs are just one example of what is happening for millions of people in many parts of Asia, especially India and China.

But the amount of fossil fuels used by the Zhangs is still small compared with the energy used by families in the USA and UK. This table shows the amount of carbon emitted (given out) in carbon dioxide by different countries from burning fossil fuels.

| Country | Population in 2006 in millions (approx.) | Carbon emissions in 2006 in millions of tonnes (approx.) |
|---------|------------------------------------------|-----------------------------------------------------------|
| UK | 60 | 160 |
| USA | 300 | 1600 |
| China | 1300 | 1400 |
| India | 1100 | 300 |

 **8** Which country had the highest carbon emissions in 2006?

**9** Find the average amount of carbon produced per person in each country. (Divide the country's emissions by its population.)

**10** How much carbon would be produced by China if it emitted the same average amount per person as the USA?

▲ Pollution over China.

## Planning for the future

The world's population is also growing so we will need to make more electricity as well as waste less energy. If we find cleaner, renewable sources of energy we may be able to support (**sustain**) these growing energy needs and protect the Earth's climate as well. We would have **sustainable** energy use.

 **11** Describe **three** cleaner ways of making electricity.

**12** Write a leaflet to persuade the government to find cleaner and renewable energy sources.

**Get this**

- Burning fossil fuels increases global warming.
- Energy use is increasing. We must:
  - waste less
  - use cleaner, renewable sources.

## Lost in space?

**Gravity** is one of the most important forces. It keeps our feet firmly on the ground.

▲ Where would we be without gravity?

▲ They don't fall off in Australia!

**1** Is gravity a pushing or pulling force?

**2** Is gravity the same everywhere in the world?

**3** People in Australia are upside down compared to us, but they don't fall off! What does this tell us about gravity?

## What do forces do?

Forces stretch and squash; pull and push; twist and bend; speed things up and slow things down; start things moving and stop things moving; change the direction in which something moves.

When scientists draw diagrams, they show the direction of forces with arrows.

**4** What does the arrow show the woman is doing to the car?

**5** What do the arrows show the man is doing to the bar?

**6** What does the arrow show the girl is doing to the sledge?

The length and thickness of the arrow shows the size of the force.

**7** How do you know that the force to pull the teddy bear is less than that used by the tug-of-war team?

## Friction!

Some forces try to stop things moving. There is **friction** between car tyres and the road. This force has to be overcome before the car moves.

There is more friction between rough surfaces than smooth ones. It's easier for the girl with the sledge to pull it across snow than gravel.

Air resists you as you push forwards through it. Water resists you even more; treacle is worse still.

Friction can be useful – it provides the grip for your tyres on the road. Air resistance slows you down when you jump out of an aeroplane and open a parachute.

## Streamlining

What do a dolphin and this aeroplane have in common?

They are **streamlined.** This allows them to move more easily through water or air. Pointed, thin, smooth shapes are more streamlined than flat, thick, rough shapes.

## Measuring forces

Forces are measured in newtons using a newton meter.

**10** What do you measure in newtons; mass, weight or both?

Force meters use the regular extension of a spring to measure force or weight.

**8** You drop a marble into a long tube containing water. It takes 3 seconds to fall to the bottom. How long will it take if the same tube is filled with treacle? Choose your answer from these options:

less than 3 seconds

exactly 3 seconds

more than 3 seconds

**9** Why do you fall slower once you have opened your parachute?

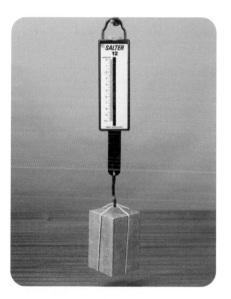

▲ A newton meter measures weight.

### Summing up

**11** Why do we make aeroplanes streamlined?

**12** The picture shows a girl diving into a pool. Why does she have her hands together above her head when she dives?

## Get this

- Gravity keeps us on the Earth.
- Forces can change the way things move.
- Friction opposes movement.
- We measure forces in newtons.

127

### Lifting!

Forces make things move.

When Erika Yamasaki won a bronze medal for Australia in the 2006 Commonwealth games, she used a lot of force. She pushed up with a bigger force than the weights pushed down. That is why the weights moved upwards. The weights she lifted were 840 N so she applied a force of more than 840 N.

The forces were **unbalanced**.

Remember; the bigger arrow on the diagram shows which force is largest.

### Learn about
- Balanced and unbalanced forces
- Speeding up and slowing down

▲ Erika Yamasaki lifts 840 N.

**1** What force pulls the weights down?

Benjamin Turner held the weights above his head.

He pushed up with the same force as the weights pushed down. That is why the weights did not move.

The forces were **balanced**.

**2** Why are the arrows on the picture the same size?

**3** If the weights weigh 1660 N what force is the weightlifter using to push them up?

After the weightlifters have finished their lift, they lower the weights back to the ground.

**4** Now which force is biggest; the weights or the force the weightlifter puts on the weights?

▲ Benjamin Turner has lifted 1660 N.

### Speeding up and slowing down

A moving car has forces acting on it.

The **thrust** of the engine provides the forward force. **Air resistance** and friction provide the backward forces.

Air resistance is a force between the car and the air surrounding the car. It depends on lots of things including the shape of the car.

Friction is the force between the tyres and the road surface.

thrust    air resistance
friction        friction

▲ Horizontal forces on a moving car.

How a car moves depends on the size of the forces.

If the thrust is larger than the air resistance and friction, the car speeds up. The car **accelerates**. When you press your foot harder onto the accelerator pedal, the engine produces more thrust.

▲ Speeds up.

If the air resistance and friction are larger than the thrust, the car slows down.

If the air resistance and friction are equal to the thrust, the car moves at a constant speed.

▲ Slows down.

**5** The thrust from a car is 2000 N. The air resistance and friction together total 680 N. What happens to the speed of the car?

**6** The thrust from a car is 590 N. The air resistance and friction together total 680 N. What happens to the speed of the car?

**7** The thrust from a car is 1600 N. The friction is 500 N. What is the size of the air resistance if the car is travelling at a constant speed?

**8** Suggest **one** other force that is acting on the car all the time.

▲ Constant speed.

## All at once

Forces act in all directions. The plane is flying at a constant speed and in level flight.

The air resistance and thrust are balanced.

The lift and weight are balanced.

**9** The aircraft slows down as it comes in to land. Which force is bigger – thrust or air resistance?

▶ Forces on an aircraft.

## Summing up

**10** What force causes a moving object to speed up?

**11** What forces cause a moving object to slow down?

**12** What happens to the speed of a car when the driver pushes down on the accelerator?

**13** Why does a bicycle slow down when the rider stops peddling?

## Get this

● Balanced forces:
  ● keep things moving at a constant speed
  ● keep things still.
● Unbalanced forces:
  ● speed things up
  ● slow things down.

### Getting rid of friction

Friction is a force that opposes movement. There is friction when an object moves over a solid surface. There is also friction when something moves through a liquid or gas.

It can be a nuisance – golfers and pool players know just how much of a nuisance this is when the ball rolls towards the hole or pocket but stops short.

▲ The 8 ball stops short of the pocket.

**Learn about**
- Where friction is a nuisance
- Why friction is useful
- How to reduce friction
- Streamlining

Friction between solid surfaces depends on the types of surfaces that are touching and the force pushing them together.

Smooth or polished surfaces have less friction. In a ski jump, the skiers get faster as they ski down the ramp and shoot off the rise at the end into the air. This skier has got stuck at the bottom of the ramp.

**1** Look at the picture. Why should the ski jumper polish his skis?

**2** Apart from polishing his skis, how else could the skier have made a successful jump?

▲ Friction can sometimes be a nuisance.

When moving surfaces rub together, the friction between them makes them hot and wears them away faster. The wheel bearings in cycles, skates and skateboards must be lubricated regularly to reduce friction.

**3** What should you use to lubricate wheel bearings on a skateboard?

◄ The skateboard wheels need to turn smoothly.

### How do we know... about reducing friction?

Rub your hands together, gently at first then harder and faster. Now do the same as you wash your hands with soap.

**4** What difference do you notice?

## Vital friction

But friction is also vital. The friction between tyres and the road makes this cycle move along.

Friction between the wheels and the brake blocks make the wheel stop turning.

 **5** Describe what would happen if you tried to ride your bike on an ice rink.

▶ The cyclist needs friction to help him move.

## How do we know... when things are streamlined?

Friction on things moving through air or water can be reduced by using streamlined shapes.

We can use a wind tunnel to see how streamlined a shape is. A stream of smoke passes over the car as air blows from front to back. The smoke shows how easily the air flows over the front and top of the car. If it stays close to the car and follows its shape the car is streamlined and air resistance is small. This means the engine needs less thrust and therefore uses less fuel.

Lorries have roof fairings to improve their streamlining.

▲ Smoke shows the path of the air in a wind tunnel.

**6** How can you tell from the air flow in the diagram that the lorry is not well streamlined?

**7** How does improving streamlining on lorries lead to cheaper food in

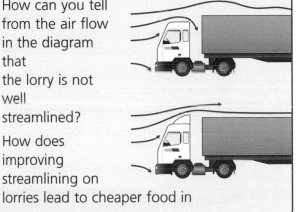

## Summing up

**8** Why does a marble roll further on a tiled floor than on carpet?

**9** Trains are often delayed because of 'leaves on the line'. Many trains are able to automatically apply sand onto the track when wheel spin occurs. What effect does the sand have?

**10** What does a free-fall parachutist do if she wants to reduce the speed at which she is free-falling?

**11** Why do the brakes of a car often smell hot after the car has gone down a very steep hill?

### Get this

- Friction comes from solids, air or water.
- It can be useful for grip or slowing things down.
- It can be reduced by lubrication or streamlining.

### What is weight?

**Weight** is the pull of gravity on an object. It is a force so it is measured in **newtons**. **Mass** is the amount of matter an object contains and does not change. It is measured in **kilograms** (kg).

100 N

On Earth, a mass of 1 kg weighs 10 N. The Moon's gravity is only one sixth as strong as Earth's, so a 600 N student on Earth only weighs 100 N on the Moon.

**1**  What is the weight of a 25 kg mass on Earth?

**2**  What is your weight on Earth?

So if want to lose weight, but not give up the chocolate, go to the Moon! You won't look any thinner and your mass will be the same but you will lose weight.

### Floaters and sinkers!

You don't have to go to the Moon to appear to lose weight. Ever wondered why you feel lighter when you're swimming? When you float, the water is pushing you up (upthrust) with the same force as your weight pulls you down. The forces are balanced and you feel weightless.

### Learn about
- Weight and mass
- Floating and sinking
- Stretching
- Reading graphs

### Brainache

**Q** So why do my bathroom scales show my weight in kilograms?

**A** Scientifically they are wrong. They should show your weight in newtons.

weight ↓ ↑ upthrust

### How do we know... that water has upthrust?

When an object that weighs 10 N in the air is weighed underwater, it appears to weigh only 6 N.

**3**  What is the upthrust of the water?

**4**  Which force is biggest here – weight or upthrust?

**5**  Draw a force diagram showing the forces on this object when it is in the water.

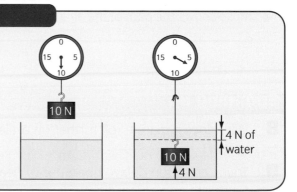
10 N
10 N
4 N
4 N of water

Imagine holding a table tennis ball at the bottom of a beaker. When you let go, the ball rises to the surface.

**6**  Which force is biggest on the ball – the weight or the upthrust, or neither?

**7**  How much does the ball appear to weigh when it floats?

## Bungee!

Bungee rope contains a large number of **elastic** cords bound together.

The rope stretches when there is a force on it. The rope returns to its original length when the force is removed. The amount of stretch is called the **extension**.

▲ Only the bungee rope to hold him.

 **8** A bungee rope is 30 m long at the start of a jump. It stretches to 33 m during the jump. What is the extension of the rope during the jump?

**9** How long is the rope after the jump?

The table and graph show how far a bungee rope stretches when weights are added.

 **10** What is the extension on the bungee rope if 500 N is added?

**11** Finish this sentence. For every 100 N added, the extension …

| Weight added (N) | Extension (m) |
|---|---|
| 0 | 0.0 |
| 200 | 0.8 |
| 400 | 1.6 |
| 600 | 2.4 |
| 800 | 3.2 |
| 1 000 | 4.0 |

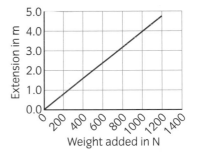

## Springing back

The same thing happens to springs. When you bounce on a trampoline the springs stretch and then return to their normal length. This pushes you back up into the air. But if your weight is too heavy the springs will either snap or be permanently stretched.

## Get this

- Weight is the pull of gravity on an object.
- Upthrust from water acts against the pull of gravity so that objects seem lighter.
- Objects float if the upthrust is greater than the weight.
- When forces stretch elastic materials or springs you can work out and predict their extension.

## Summing up

**12** Why is it important to know how much a bungee rope extends when it has different weights on the end?

**13** Why do trampolines have weight limits?

**14** The QM2 weighs 1500 million newtons. What is the upthrust on the ship?

### What is speed?

**Speed** is a measure of how fast something is moving. The unit of speed is **metres per second** (m/s) or **kilometres per hour** (km/h). In everyday life in the UK we still use miles per hour (mph) on the roads.

To find out the speed of a moving object, we measure the time it takes to travel between two points a set distance apart.

In the early days of motoring, the police did just that. Two police officers measured the distance between two points. One officer waved his hand as the car passed the first point.

The second officer started a stop watch and stopped it when the car passed the second point. If the car was speeding, the second officer had time to stop the car. Don't forget, cars only travelled at speeds up to about 20 miles per hour.

Speed is calculated as distance divided by time:

$$\text{speed} = \frac{\text{distance}}{\text{time}}$$

So if a runner runs 100 metres in 10 seconds his speed is:

$$\frac{100}{10} = 10 \text{ metres per second}$$

**1** Robyn runs 200 m in 40 s. Calculate her speed in metres per second.

**2** David drives 150 km in 2 hours. Calculate his speed in kilometres per hour.

Today, SPECS speed cameras catch speeding drivers automatically. The cameras are placed a known distance apart. This is usually about a mile. They read a car number plate as it passes the first camera and record the exact time. The same thing happens as the car passes the second camera. A computer works out if the car is speeding and the car owner receives a letter and fine.

### Distance–time graphs

You can show speed on a graph. This graph shows how something moves at a constant speed. The distance travelled each second is the same. The graph is a straight line.

**3** Look at the graph. How far does the object travel in 10 s?

**4** How fast is the object travelling during the whole journey?

### Learn about
- How forces affect speed
- Showing speed on a graph
- How to calculate speed

▲ An early police speed check.

▲ SPECS speed cameras.

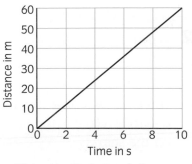

▲ Distance–time graph for constant speed.

## Average speed

In real life, few things move at constant speed. Most objects or people slow down and speed up during a journey.

We normally refer to average speed as:

$$\text{average speed} = \frac{\text{total distance travelled}}{\text{total time taken}}$$

Lucy lives 4000 m (4 km) from school. She walks to school and the journey usually takes her 40 minutes. The graph shows her journey one morning.

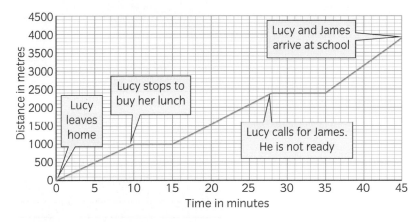

▲ Distance–time graph for a typical journey to school.

**5**  How far is Lucy from home when she stops to buy her lunch?

**6**  How long did it take her to buy her lunch?

**7**  How far from Lucy does James live?

**8**  How long did it take Lucy to reach James' house?

**9**  How long did James keep Lucy waiting?

**10**  It usually takes Lucy 40 minutes to walk to school. Calculate her average speed in metres per minute.

### Summing up

**11**  What two quantities do you need to know to calculate speed?

**12**  A snail can move 1 cm in 30 seconds. If it moves at a constant speed, how long will it take to travel 10 cm?

**13**  The forces on a train are balanced. Is the train:

A speeding up

B slowing down

C travelling at a constant speed?

## Brainache

**Q** What happened to Lucy's speed after she left James' house? How does the graph show this?

**A** Lucy had to walk quicker. The slope of the graph is steeper.

## Get this

- Speed is the distance travelled in a certain time.
- In science, speed is usually measured in metres per second.
- Graphs can show how speed varies during a journey.

## Pushes and pulls

Ben is a fitness instructor at a Leisure Centre. It's his job to make sure people use the equipment correctly and do not try to use more force than they should.

Forces can pull and push.

**Learn about**
- Forces in the fitness centre
- Forces on the building site

**?** **1** What type of force is used in doing press-ups or using the bench press?

**2** What force would be needed to lift 400 N weights?

**3** What type of force is used with a rowing machine?

## Stretches and twists

Forces can stretch and squash; twist and bend.

**?** **4** What type of force is used with a bullworker?

**5** Sam's right arm exerts a force of 200 N on the bullworker. What force does his left arm exert on the bullworker?

## More friction

A treadmill works because of friction. Force is needed to start the treadmill working and to keep it moving. More force is needed to run faster. The surface on which you run is a continuous belt. As you run forwards, friction between your feet and the surface causes the belt to go backwards.

Friction between the belt and the rollers turns the rollers. The front roller is linked to a speedometer to show how fast you are running.

**?** **6** What type of surface is the belt on a treadmill?

**7** We often lubricate the moving parts of a machine. Explain why you would not lubricate the surface of the rollers on a treadmill.

**8** The belt on the treadmill stretches a bit and does not grip the rollers properly. How might this affect the reading on the speedometer?

## Only tools and forces!

Bob is the foreman on a building site. He has to make sure that the forces in the building are safe.

When builders are working, they use forces and Bob has to make sure they use the right forces.

## Plumber's forces

A plumber uses spanners when she joins lengths of pipe together.

She secures the pipe to the wall with a pipe clip.

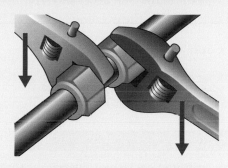

**9** What forces are at work as the clip holds the pipe?

**10** A plumber is using two spanners as shown in the diagram to tighten a nut. When the nut is as tight as it can be, what can you say about the sizes of the forces on each spanner?

## Carpenter's forces

A carpenter joins wood together with nails, screws or bolts.

**11** What forces is a carpenter using?

**12** Draw diagrams to show a hammer and a screwdriver in use. Add arrows to show the directions of any forces.

## Forces at home

**13** We use different forces in doing a simple task such as preparing breakfast:

- **push** down on toaster lever; switch on kettle
- **pull** open drawer; knife across butter
- **twist** lid off marmalade
- **squash** fresh orange.

List other examples of how you use forces when preparing breakfast.

### Get this

- Forces push and pull; stretch and squash; twist and bend.
- Many forces are used in a fitness centre.
- Many forces are used on a building site.
- We all use forces – all day, every day.

**Learn about**
- Day time and night time
- The seasons of the year

### Sunrise, sunset

Imagine an early caveman watching the Sun as it passes across the sky. He may have thought there was a new Sun every day.

He then probably worked out that it was the same Sun every day. He saw the Sun rise in the morning and disappear in the evening.

We still call it sunrise and sunset. The Sun 'rises' in the east.

**1** In which direction does the Sun set?

Today, we know that the Earth orbits the sun.

### Night and day

Night and day are caused by the Earth spinning on its axis every 24 hours. This gives us day time and night time. Half of the Earth is always in darkness. We can see this clearly from space.

**2** What is orbiting the Earth and taking pictures of it?

**3** Where is the Sun in this photo: right or left?

### How long is a day?

The Earth's axis is tilted. It is also orbiting the Sun as it spins on its axis.

When it's summer in the United Kingdom, the northern hemisphere is tilted towards the Sun and the southern hemisphere is tilted away. This means that we spend longer facing the Sun as the Earth spins and we have longer day time than night time. It's the other way round in winter.

**4** It is June in Australia. Which is longest, day or night?

Summer at the North Pole means constant daylight.

**5** What is the length of night time at the North Pole in winter?

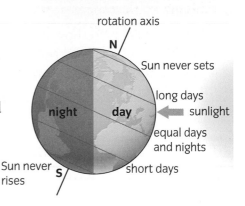

rotation axis

N

Sun never sets

long days

← sunlight

night    day

equal days and nights

Sun never rises

S

short days

During the year, the hours of daylight constantly change. The graph shows how the hours of daylight change in Birmingham.

**6** Which month has the longest hours of daylight?

**7** In which months are there 12 hours of daylight and 12 hours of night time?

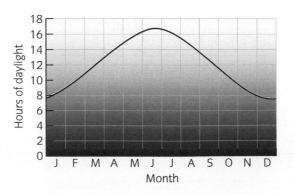

## The four seasons

The Earth's tilt on its axis as it orbits the Sun also gives us the four seasons in the northern and southern hemispheres. The long hours of sunlight in summer warm the Earth. The short hours of sunlight in winter means it gets colder.

**8** How long does it take for the Earth to orbit the Sun?

**9** Why is it warmer in summer and colder in winter in the UK?

**10** Why do people in Australia flock to the beach for their Christmas dinner?

**March**
spring in the north and autumn in the south

Sun

**June**
summer in the north and winter in the south

**December**
winter in the north and summer in the south

**September**
autumn in the north and spring in the south

## A question of angles

The rays from the Sun are also more concentrated in summer because they strike the surface of the Earth full on, almost at a right angle (see diagram). This makes the heat from them very concentrated.

In winter the rays strike the Earth at a more oblique angle. You can see from the diagram that this means they are 'spread out' when over a wider area. This is why it is colder in winter.

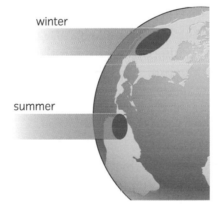

winter

summer

**11** Why does the Sun look higher in the sky in summer than in winter?

### Summing up

**12** Which part of the Earth always has about 12 hours of daylight?

**13** Why is there no winter or summer at the equator?

**14** Which month has the shortest hours of daylight in New Zealand?

**15** The shortest length of daylight in England is on December 21st. When is the shortest length of daylight in Canada?

## Get this

- The Earth spins on its axis every 24 hours and this causes day and night.
- The Earth's axis is tilted. This explains:
  - the different lengths of daylight
  - the seasons as the Earth orbits the sun.

139

**Learn about**
- The planets in our Solar System
- Some of the other bodies orbiting the Sun

### The planets

There are eight **planets** in our **Solar System**. They all orbit the **Sun** in roughly circular paths.

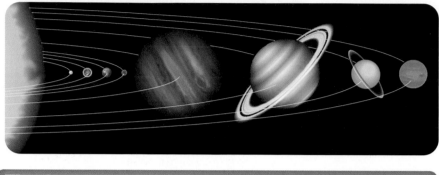

### The inner planets

Mercury, Venus, Earth and Mars are known as the inner planets.

Mercury is closest to the Sun. It is covered with **craters** and looks like our own Moon. During the day it can be as hot as 430 °C. At night, it cools to −170 °C.

Venus reflects a lot of light from the Sun and can often be seen in the evening, just after sunset. Its atmosphere is carbon dioxide, which is a **greenhouse gas** and traps heat from the Sun. This means the planet is even hotter than Mercury. Sometimes Venus is called the Evening Star but it is not a star.

**? 1** Stars produce their own light. How can we see other objects such as planets and moons?

Mars is the only other planet that may once have had life. The planet is now too cold for there to be any liquid water, but there is evidence that water flowed on its surface billions of years ago. There may still be water underground.

**? 2** Why do scientists think there may have been life on Mars a long time ago?

### The asteroids

Between Mars and Jupiter is the **asteroid belt**. Asteroids are lumps of rock left over from when the Solar System was formed. The largest asteroid is called Ceres. It is about 1000 km across.

Some asteroids have orbits that pass close to Earth and some even hit it. About 65 million years ago, an asteroid landed in Mexico. We think it caused the climate to change. This meant that dinosaurs could not survive.

▲ Mercury

▲ Venus

▲ Mars rover on the planet's surface.

▲ Asteroid Ida with its moon Dactyl.

**3** Scientists are constantly looking at the paths of asteroids to see if they come close to Earth. Why do they do this?

## The outer planets

The four outer planets are sometimes called the gas giants. They are made mainly of gas and liquid. Temperatures on these planets vary from −150 °C to −220 °C.

All the gas giants are surrounded by one or more rings. Saturn has thousands. The particles in Saturn's rings may be as small as a grain of sugar or as large as a house. The particles may have come from bits of comets or asteroids that broke up before reaching the planet.

▲ Saturn

## Planet fact file

**4** How far is Jupiter from the Sun?

**5** Which is the largest planet?

**6** Roughly how much larger is the Sun than the largest planet?

| Name | Distance from Sun in million km | Diameter in km |
|------|---------------------------------|----------------|
| Sun | | 1 391 000 |
| Mercury | 58 | 4 879 |
| Venus | 108 | 12 104 |
| Earth | 150 | 12 756 |
| Mars | 228 | 6 787 |
| Jupiter | 778 | 142 800 |
| Saturn | 1 427 | 120 660 |
| Uranus | 2 871 | 51 118 |
| Neptune | 4 498 | 49 528 |

## How do we picture... the scale of the Solar System?

It's very difficult to imagine the enormous distances in the Solar System. We can make a scale model to get some idea. If you divide all the distances from the Sun by a million and then again by 100 000 you have some distances you can work with:

'Neptune' in the model is 44.98 metres from the Sun. 'Mercury' is 58 cm away.

**7** Using this scale, work out how far away from the Sun each planet will be.

## Summing up

**8** Why do you think planets are called 'inner' and 'outer'?

**9** Why is the temperature on Venus hotter than on Mercury?

**10** Why is it difficult to draw the Solar System to the right scale in a book?

**Get this**

- There are eight planets and an asteroid belt in our Solar System.
- Models help us to understand the size of the Solar System.

### Orbits

The gravitational force of the Sun keeps the planets moving in a circle. The Moon orbits the Earth.

**1**  What is the source of the gravitational pull on the Moon?

As the Moon orbits the Earth, it spins on its axis. The time it takes to spin is exactly the same as the time it takes to orbit. This means that the Moon always shows the same face towards Earth.

Most planets have moons orbiting them.

| Planet | Number of known moons in May 2007 | Time to orbit Sun in Earth days |
|--------|-----------------------------------|--------------------------------|
| Mercury | 0 | 88.0 |
| Venus | 0 | 224.7 |
| Earth | 1 | 365.26 |
| Mars | 2 | 687.0 |
| Jupiter | 62 | 4 332 |
| Saturn | 59 | 10 761 |
| Uranus | 27 | 30 685 |
| Neptune | 13 | 60 190 |

**2**  In 1997, Jupiter only had 16 known moons. Suggest why many more are known today.

Planets travel at different speeds around the Sun.

**3**  Suggest **two** reasons why Neptune takes longer than Saturn to orbit the Sun.

**4**  Which planet has a year length which is approximately 3 Earth months?

### Phases of the Moon

As the Moon orbits the Earth, we see it at slightly different angles in the sky. It also appears to change its shape. This is because the Moon reflects light from the Sun. When it is at a different angle we see a different part of the Moon being lit.

The diagram shows the phases of the Moon as we see it from Earth.

**5**  What fraction of the Moon can be seen from Earth when the Moon is at its third quarter phase?

**6**  How do we describe the Moon when it cannot be seen from Earth?

## Learn about
- Different orbits
- Eclipses of the Sun and Moon

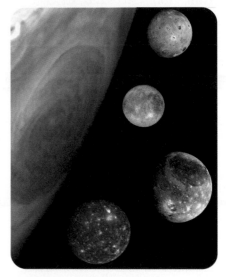

▲ Jupiter and its moons.

### Brainache

**Q** What pattern describes how the speed of a planet changes with distance from the Sun?

**A** The greater the distance, the slower the planet's speed.

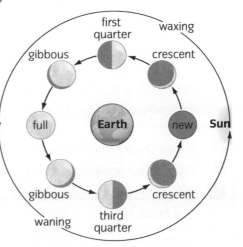

▲ The Moon as seen from Earth.

## Eclipses

Sometimes, as the Earth orbits the Sun, the Moon comes between the Sun and the Earth. When this happens there is an eclipse of the Sun. The Moon blocks out the light from the Sun.

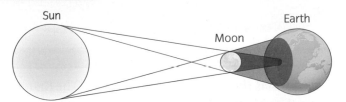

 **7** An eclipse of the Sun can only be seen from certain points on Earth. Suggest why.

An eclipse of the Moon happens when the Sun, Earth and Moon are in a perfectly straight line. The Moon passes into the shadow of the Earth.

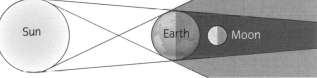

Eclipses are more common than most people think. A total eclipse of the Sun happens about every 18 months, but it is rarely visible from land. Total eclipses of the Moon are less frequent but are visible from all parts of the hemisphere. This is why they seem to be more common.

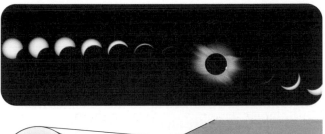

## Comets

Comets also orbit the Sun. Their orbit is not circular like the planets.

Their orbits pass close to the Sun and stretch well beyond Neptune. Comets are made of dust particles frozen in ice. When they pass close to the Sun, the ice melts and a tail is formed. The tail is blown by a Solar wind. It always points away from the Sun.

**8** What makes a comet's tail visible?

▲ Comet Hale Bopp

### Summing up

**9** What is the source of the gravitational pull on the planets?

**10** What is the difference between a waxing and a waning Moon?

**11** Suggest why birds stop singing and return to their nests during a total eclipse of the Sun.

**12** What is a comet doing if it is moving with its tail in front of it?

### Get this

- Gravity keeps bodies in orbit.
- The Moon appears differently every night as it orbits the Earth.
- Comets have elliptical orbits.

**Learn about**
- Uses of artificial satellites
- Types of orbit

### Artificial satellites

The Moon is the Earth's natural satellite. The first artificial satellite, Sputnik 1, was launched on October 4 1957 by the Russians. In the 50 years since then, there have been over 5000 satellites launched into space. There are still 850 in orbit.

At least 500 of the satellites in orbit are used for communications. Worldwide telephone calls, internet signals and television programmes are beamed up to the satellites from one part of the world. The signals are amplified and then beamed down to receivers in other parts of the world.

These satellites take 24 hours to orbit and the Earth spins on its axis every 24 hours. This means that the satellite sending a signal to your house appears to be in the same place in the sky. After all, you don't want to have to keep moving your satellite dish.

**1** Communications satellites travel in the same direction as the Earth spins. Why?

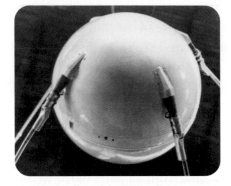

That should be OK for another 10 minutes.

### Weather

Some satellites are used for monitoring and taking pictures of weather systems on Earth. They are often in **polar orbits**. This means they orbit along a North-South direction. As the Earth spins beneath them, they have a view of the whole Earth surface during a day.

**2** Weather satellites orbit much closer to Earth than other types. Why do weather satellites orbit close to Earth?

**3** Weather satellites only take 100 minutes to orbit the Earth. Why do they take less time than communications satellites?

### GPS and satellite navigation

About 50 satellites are used by **Global Positioning Systems** (GPS). These help people find their position on the Earth.

GPS has been used by ships for a long time but can now be used by everyone in a car with a **satellite navigation** device (satnav for short).

From any point on the Earth's surface, at least four satellites are overhead emitting signals. Three satellites can identify a particular position. The fourth is used to confirm it.

A GPS receiver on the Earth calculates the distance to each satellite. It can then identify the precise location on a map stored in the receiver's memory.

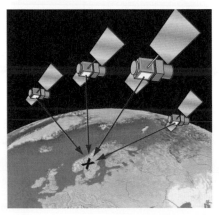

GPS is also used by the military to target missiles more accurately. It can be used to synchronise clocks and track animals and convicted criminals by electronic tagging.

**4** If someone is tracked by electronic tagging where is the GPS receiver?

**5** Look at the newspaper article. Suggest why satnav sometimes sends people along unsuitable routes.

## THE DAILY MAIL
16th March 2007

### £96,000 Merc written off as satnav leads woman astray

THE DRIVER, the latest of many to be led astray by satnavs, was on her way to a christening party in Leicestershire when she was sent down a winding track usually used only by farmers in their 4x4s.

Although the track is signposted as 'unsuitable for motor vehicles', the driver carried on and found herself at a ford in the village of Sheepy Magna. Still accepting what the satnav told her, she set out to cross the ford, but it was swollen after days of heavy rain.

## Space station

Some people live in space. The **International Space Station** (ISS) orbits the Earth every 90 minutes at a height of 340 km. It was assembled in space by **astronauts** and is so large that you can see its reflection as it moves across the night sky. While they are on board, the astronauts do a lot of different experiments.

**6** Use the Internet to find out when and where you can see the International Space Station.

## Staying up there

How does a satellite stay in orbit? Isaac Newton worked out how over 300 years ago. He said that when you fire a cannon ball, gravity pulls it back down to Earth. But the Earth is not flat. If you could fire a cannon with so much force that as the ball fell back to Earth, the curved surface of the Earth was dropping away beneath it, the ball would just carry on going.

Satellites in orbit are being attracted towards the Earth, just like the cannon ball. They are put into orbit from a spacecraft. As they are released, they are travelling at the same speed as the spacecraft.

Satellites are travelling so fast that they are falling to Earth at the same rate as the Earth is dropping away from them. They stay in orbit. The same principle applies to moons in orbit around planets and planets in orbit around stars.

## Summing up

**7** What does 'satellite' mean?

**8** Why was the International Space Station assembled in space and not launched as a complete satellite?

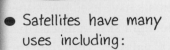

### Get this

- Satellites have many uses including:
  - communication
  - weather forecasting
  - positioning.
- Gravity keeps satellites in orbit.

### Learn about
- What's in the Universe
- How the Universe started

### What's out there?

Our star (the Sun) and its Solar System are part of a **galaxy** called the **Milky Way**. A galaxy is a number of stars grouped together.

We cannot count the number of stars in the Milky Way, but a rough estimate is five hundred thousand million. That's 500 000 000 000 stars.

The Milky Way is a spiral galaxy. Its total mass is nearly two million million times the mass of our Sun. That's quite heavy. We're about two thirds of the way out from the centre of the galaxy.

 **1** What's the difference between a solar system and a galaxy?.

The nearest star to our Sun is called **Proxima Centauri** (the bright red star in the centre of this photo). It was discovered in 1915 and is only visible with a telescope. The star is usually only visible from the southern hemisphere.

There are more than one hundred thousand million galaxies in the Universe. The nearest galaxy to our own is called **Andromeda**. It is a very long way away. In fact, the distance from our Sun to the centre of Andromeda is 24 000 000 000 000 000 000 km. Andromeda is also a spiral galaxy.

The best estimates suggest that there are at least 70 thousand million million million (70 sextillion or 70 000 000 000 000 000 000 000) stars in the Universe. Try counting them!

How do we know how many stars there are in the Universe?

**2** You are given a book with about 1000 pages in it. How could you estimate the number of words in the book?

Scientists have used telescopes to look at a strip of the night sky. In that strip, they found about 10 000 galaxies. They then measured the brightness of each galaxy. From this they worked out how many stars there were in each galaxy and then in the strip of sky.

**Astronomers say there are more stars than grains of sand in all of Earth's deserts and beaches.**

That number was then multiplied by the number of similar sized strips needed to cover the entire sky.

## Space distances

What is the fastest thing you can think of?

 **3** How long do you think it would it take this aircraft to fly to the Sun – less than 10 minutes, about 3 weeks or over 4 years?

Light travels at the fantastic speed of 300 000 kilometres per second (300 000 000 metres per second). But the Sun's light still takes just over 8 minutes to reach us.

Distances in space are so large it gets difficult to use kilometres as a unit of distance. Instead scientists who study space (**astronomers**) use light years. A light year is the distance light travels in one year.

Light from Proxima Centauri takes 4.22 years to reach us, so it's 4.22 light years away. Andromeda is 2.5 million light years away. Light from some stars takes over twelve billion years to reach us.

**4** How long does it take light from Andromeda to reach us?

**5** Work out how far light travels in a year.

## The Big Bang

Scientists think that the Universe started forming in a **Big Bang** fifteen billion years ago. Before Big Bang, the matter that made the whole Universe was packed into a space no bigger than an atom. It exploded and the explosion was so huge that it is still expanding outwards.

After one billion years, galaxies started to form. Around 9 billion years later our Sun and Solar System formed, including the Earth.

There was no life on Earth to start with. Dinosaurs appeared 245 million years ago but they died out around 60 million years ago. The first human only appeared between 2 and 3 million years ago.

**6** Draw a timeline to show the key events from Big Bang to humans appearing.

**7** The picture shows a caveman with his pet dinosaur. Why is this picture scientifically incorrect?

## Summing up

**8** Why do astronomers use light years instead of kilometres as a unit of distance?

**9** Stars eventually burn out and die. Some of the stars we see in the sky now may not exist anymore. Why?

**10** Amy wants to be an astronaut when she's older. She wants to travel to other galaxies in the Universe. Explain why she won't be able to do this.

**Get this**

- Our galaxy is the Milky Way.
- Galaxies are large collections of stars.
- Distances in space are so big we measure them in light years.
- We think the Universe started with the Big Bang.

**Learn about**
- How we look at objects in space
- Recent discoveries in space

### The first astronomers

Humans have always gazed into the sky. They looked at the patterns of stars. Some patterns looked like animals. The patterns were given names such as Leo or Cancer. We still use these names today.

People could see the planets as far as Saturn. But no one could see any further until the invention of the telescope around 400 years ago.

The first person to use a telescope to look into space was Galileo. He was a scientist in Italy. He saw craters on the Moon and discovered four of Jupiter's moons in 1610. He also discovered Saturn's rings.

Galileo is most famous for arguing that the Earth revolved around the Sun. Most people believed that the Earth stayed still and everything else revolved around it. Galileo was even sent to prison for saying this.

### Opening up space

The type of telescope Galileo used was a refracting telescope. Refracting telescopes focus light from a distant object using two glass lenses. But they are very limited. Galileo's telescopes could only make things look 30 times larger.

In 1704 Isaac Newton invented a new type of telescope which used a large curved mirror to gather light from a distant object and reflect it onto a flat mirror which focuses the image through the eyepiece. Large mirrors are easier to make than large lenses and these telescopes could magnify objects a lot more.

With the new telescopes, many more discoveries were made about the Solar System and Universe. As technology in making mirrors improved, telescopes got more and more powerful. Newton's curved mirror was 15 cm wide but big telescopes used by professional astronomers today can have mirrors 6 metres wide and have to have whole buildings to house them!

Uranus was discovered in 1781 and Neptune in 1846. It was not until 1930 that Pluto was discovered and added to the list of planets.

objective lens     eyepiece lens

light     image

eyepiece    image     concave mirror

flat mirror     light

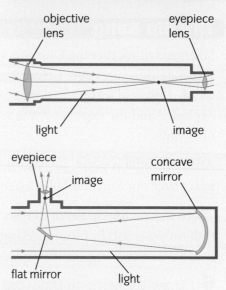

**? 1**   Why are reflecting telescopes used when a more powerful telescope is needed?

**2**   Suggest why it took until 1930 before Pluto was discovered.

## Views from space

Looking at space from Earth is a problem even with the most powerful telescopes. The atmosphere blurs images from space and scatters background light. In the past 50 years scientists have found a way round this – get out into space beyond the atmosphere!

The Hubble Space Telescope is a reflecting telescope. It was launched in 1990 and has been sending back pictures of space ever since. At a height of 569 km, it orbits above the Earth's atmosphere.

Hubble can capture pictures impossible to see from Earth such as the birth and death of stars in deep space. This photo shows a 'ring of pearls' surrounding an exploding star around 163 000 light years away.

**3** Why can the Hubble Space Telescope see things in space that we cannot see from Earth, even with very powerful telescopes?

Recently, other bodies have been discovered beyond Pluto. Some are a similar size to Pluto and so could have been added to the list of planets. But, in August 2006, astronomers agreed a new definition of a planet. Pluto got demoted to a 'dwarf planet'.

Scientists have also started sending probes to other planets which have sent back even more detail. Not all of them land; some just pass close to planets and moons to gather detailed data.

The Mars Rover has gathered a lot of information about the planet. A grinder at the end of a robotic arm can make a hole 45 mm across and 5 mm deep in the rock. The sample is analysed on board the rover and the information sent back to Earth.

## Manned exploration

So far, the Moon is the only other place in the Solar System humans have visited. The first Moon landing was on 21 July 1969.

The photo on the right was taken by Neil Armstrong. It shows Buzz Aldrin stepping onto the Moon's surface.

NASA hopes to have a permanent base on the Moon by 2024 and send astronauts to Mars after 2030. The cost could be as high as £400 billion.

**4** Who was the first man to step onto the Moon?

**5** It takes 6 months to reach Mars. What will this mean for astronauts going to Mars?

**6** Do you think it's worth the money?

### Get this

- The invention of the telescope 400 years ago changed our view of the Universe.
- Space travel and satellites have hugely expanded our knowledge of the Universe.

# Glossary

## A

**accelerate** speed up

**acid** a solution with a pH less than 7

**acidic** a substance containing acid

**air pressure** the force exerted by air particles when they collide with a surface

**air resistance** force on an object moving through the air causing it to slow down

**alkali** a solution with a pH of more than 7

**alkaline** a substance containing alkali

**ammeter** device for measuring electric current in a circuit

**ammonium chloride** a chemical called a salt made from ammonia and chlorine

**amoeba** an organism made from only one cell

**ampere (A)** unit of measurement of electric current

**amphibians** vertebrates with moist skin that can live on land and in water

**anaemic** looking pale and feeling tired – symptoms of not having enough haemoglobin in the red blood cells

**animal cell** a cell which forms part of an animal

**Andromeda** the nearest galaxy to the Milky Way

**animals** organisms that take in food to live and grow

**arthropods** animals with jointed legs and a hard outer covering

**artificial satellites** spacecraft made by people to orbit the Earth for various purposes

**asteroids** lumps of rock orbiting the Sun that were left over from when the Solar System was formed

**astronaut** someone who travels beyond the Earth's atmosphere

**astronomer** scientist who studies space

**atom** smallest particle of an element that can exist

**average** typical value – found by adding a set of values together and dividing by the number of values

**average speed** the total distance travelled in the total time taken for a complete journey

**axis**, **of Earth** the imaginary line through the Earth which it spins around

## B

**balance** equal and opposite – cancelling out

**balanced forces** when the opposite forces on an object are equal so the object does not move

**battery** two or more electrical cells joined together

**behaviour** the way animals act

**behaviour patterns** repeated ways of behaving which are inherited or learned

**Big Bang** explosion in space which we believe started the Universe

**biodiversity** the number and variety of different species of plants and animals in the world

**biomass energy** energy released by burning biomass, eg for electricity

**biometrics** identifying people using unique physical characteristics, e.g. fingerprints

**birds** vertebrates that have wings and feathers

**blood flows** through our circulatory system carrying everything needed for life

**blood groups** different types of blood

**boil** change from a liquid into a gas at the boiling point

**boiling point** temperature at which a liquid boils and changes to a gas

**bone** hard, rigid material which makes up the skeleton

**bone cells** the cells which make up bone

**brain** the part of the nervous system that controls the body

**breed** produce offspring

**burning** being on fire due to a reaction with oxygen

## C

**calcium** silver-white metallic element

**calcium carbonate** a chemical (salt) often known as limescale

**carbon** a non-metal element found as charcoal, diamond or graphite

**carbon dioxide** gas found in small amounts in the atmosphere which green plants use to make food

**carbon emissions** the carbon released into the air as part of carbon dioxide

**carbon footprint** the amount of carbon dioxide a person's activity produces in a year

**carbonates** a group of chemicals (salts) which make carbon dioxide when they react with acid (e.g. calcium carbonate)

**cell** (electrical) stores chemical energy and converts it to electrical energy in a circuit

**cell** (living) the building block living things are made from

**cell membrane** the outer part of all cells which hold in the other parts of the cell

**cell sap** fluid inside plant cells where food is stored

**cell walls** tough outer covering of plant cells which helps keep them rigid

**cellulose** tough fibres that cell walls are made of

**change of state** change from one form of matter (solid, liquid or gas) to another

**chemical energy** energy stored in fuels, food and electrical batteries

**chemical reaction** an event which creates new substances

**chemical symbols** letters which represent elements (usually one or two letters) which are understood in all languages

**chloroplasts** the green parts of plant cells which capture light energy to make food from water and carbon dioxide

**chromatogram** a record of the results of chromatography

**chromatography** a way of separating soluble substances

**cilia** tiny hair-like structures that sweep things along tubes in the body

**ciliated cells** cells with tiny hair-like cilia to sweep things along tubes in the body

**circuit** a complete pathway for an electric current to flow

**circuit diagram** a way of showing a circuit clearly using symbols

**circuit symbol** a drawing which represents a component in a circuit

**circulatory system** the organ system which sends blood to all parts of the body consisting of the heart, blood vessels and blood

**classification** sorting things into groups according to their similarities and differences

**coal** a fossil fuel formed from plants over millions of years

**combustion** burning

**comets** bodies in space made of dust particles frozen in ice which orbit the Sun

**compass** device containing a small magnet that is used for finding directions

**component** an item used in an electric circuit, e.g. a bulb

**compound** substance made of two or more elements chemically joined together

**compress** squash into a smaller space

**concentrated** a solution that contains very little water

**condense** turn from a gas into a liquid

**conifer** plant that produces seeds in cones

**conjoined twins** twins formed when a partially split embryo develops who often share vital organs

**conservation** (of energy) energy is never made or lost but is always transferred from one form to another, although they are not always forms we can use

**continuous variation** inherited features which can have any value in a wide range, e.g. height, weight

**contract** (biology) get shorter and fatter

**contract** get smaller

**core** rod of a magnetic material placed inside a solenoid to make the magnetic field of an electromagnet stronger

**correlation** a link between two things

**corrosive** destroys living tissue

**crude oil** a thick black liquid usually found underground and used to make a number of fuels e.g. petrol, diesel, and also many plastics

**current** the flow of electrical charge (electrons) around a complete circuit

**cytoplasm** the jelly-like substance inside a cell where most of its activity happens

**D**

**Dalton, John** English scientist in the 1800s whose experiments started the development of the modern theory of the atom

**day time** the period on one section of the Earth when it is facing the Sun

**density** the number of particles of matter in a certain volume

**diesel** a fuel made from crude oil

**diffuse** spread out

**diffusion** spreading out

**digestive system** the organ system which digests food in the body consisting of the mouth, stomach and intestine

**dilute** a solution that contains lots of water

**discontinuous variation** inherited features which have distinct categories, e.g. eye colour, ability to roll the tongue or not

**dissolve** when a solid mixes with a liquid to make a solution

**distance–time graphs** these show the distance travelled over a period of time, that is, speed

**distillation** a way of separating a solvent from a solution

**E**

**Earth** a rocky inner planet third in order of planets from the Sun

**eclipses** when the Sun or Moon is blocked from view on Earth (see also lunar eclipse or solar eclipse)

**egg cell** female sex cell

**ejaculate** release semen from the penis

**elastic** a material which can be stretched and return to its original length

**elastic potential energy** energy stored in an elastic object that is stretched or squashed

**electric circuit** a complete pathway for an electric current to flow

**electric current** flow of electric charge (electrons) around a complete circuit

**electrical energy** energy which makes current flow around a circuit and is changed in circuit components to other forms of energy (e.g. a bulb changes it to light energy)

**electricity** the flow of electrical energy in a current of charge (electrons)

**electromagnet** temporary magnet created using an electric current

**electron microscope** very powerful microscope that can magnify objects up to one million times

**electrons** tiny charged particles that flow through a wire and create an electric current

**element** substance consisting of atoms of only one type

**embryo** a plant or animal that is just beginning to grow from a fertilised egg before it has all its organs

**energy** needed to make things happen

**energy changes** when energy moves from one form to another (the scientific name is transfer)

**energy conservation** energy is never made or lost but is always transferred from one form to another, although they are not always forms we can use

**energy crisis** the problem caused by increasing demand for energy and fossil fuel resources of energy running out

**energy transfer** when energy moves from one form to another, e.g. from electricity to heating

**environmental variation** variation as a result of environmental influences, e.g. scars

**Equator** imaginary line round the middle of the Earth at an equal distance from both the North and South poles

**erection** when the penis swells with blood and stiffens

**estimate** do a rough calculation to get an answer close to the right one

**ethologist** a person who studies the way animals respond to their environment

**evidence** something that provides proof of a scientific theory

**expand** get bigger

**extension** the amount an object stretches

**external fertilisation** when sex cells (egg and sperm) join outside the body usually in water (e.g. in most fish and frogs)

**extinct** when every member of a species has died

**F**

**fat cells** large cells filled with oil which is an energy store for the body

**female reproductive system** the organ system in females for producing offspring, consisting in humans of ovaries, oviducts, uterus, vagina

**ferns** plants which do not have flowers or seeds and reproduce by spores

**fertile** able to produce offspring

**fertilisation** when two sex cells join to make a new organism

**fertilise** what a sperm does to an egg when they join together and become one cell

**fetus** the name for a young animal developing in the womb when it has all its organs (after 8 weeks in humans)

**fingerprints** patterns in the skin at the ends of fingers which are different in every person

**fish** water-living vertebrates with gills, scales and fins

**floating** an object floats when the upthrust from water is equal to the downwards force of the object's weight

**flowering plants** plants which produce flowers and reproduce by the seeds they make

**food** a source of chemical energy for animals

**food chains** diagrams which show what organisms eat, with arrows showing how energy flows through the chain; always start with a plant

**food energy** the chemical energy stored in food

**food webs** diagrams showing the connections between different food chains

**force meter** device used to measure forces

**forces** these act on objects and affect their movement, e.g. pushes, pulls, gravity, upthrust, friction

**fossil fuels** fuels made from the remains of animals and plants that died millions of years ago, e.g. coal, oil, natural gas

**freeze** turn from a liquid into a solid

**friction** a force that resists movement

**fuel** a store of energy which can be burned, e.g. in gas, oil, coal, petrol

**G**

**galaxy** a number of stars and the solar systems

around them grouped together

**Galileo** Italian scientist in the 1600s who first explored space with a telescope and believed that the Earth went round the Sun

**Galvani, Luigi** Italian doctor who investigated the effects of electric charges on frog muscles

**gas** a fluid with no fixed volume that takes the shape of its container

**gas pressure** the force exerted by gas particles when they collide with a surface

**genes** parts of a DNA molecule that control a certain characteristic

**geothermal energy** energy contained in the hot rocks beneath the Earth's surface

**Global Positioning System (GPS)** system which pinpoints the position of something using signals from a satellite

**global warming** gradual increase in the Earth's average temperature

**glucose** the sugar made in photosynthesis

**gravitational potential energy** energy stored in an object because of its height above the ground

**gravity** the force exerted on objects by masses such as planets, moons and the Sun

**greenhouse gases** gases that contribute to global warming, e.g. carbon dioxide

## H

**haemoglobin** red chemical in red blood cells which carries oxygen

**hazard symbols** warning symbols on some chemicals which show what harm they might cause if not handled properly

**heat energy** possessed by hot objects, also called thermal energy

*Homo floresiensis* a newly discovered species of human

*Homo neanderthalensis* Neanderthal man – an extinct species of human

*Homo sapiens* the name of our species

**Hooke, Robert** English scientist in the 1600s who discovered cells

**hormones** special chemicals that are released into the bloodstream and affect certain organs or parts of the body

**Hubble space telescope** a telescope orbiting the Earth and sending back clear pictures of space

**humans** the species of vertebrate, warm blooded mammals to which we belong

**Human Fertilisation and Embryology Authority (HFEA)** the authority in the UK which sets guidelines for IVF

**hydrocarbon** a compound containing hydrogen and carbon only

**hydrochloric acid** a strong acid

**hydroelectric power** electricity generated using the energy of water falling downhill

**hydrogen** a non-metal element that exists as a gas at everyday temperatures

**hypothesis** an idea, theory or interpretation of a situation which has not been tested

## I

**Ibn Hayyan, Jabir** scientist in the 800s working in an area now in Iraq who made many important discoveries in chemistry, including acids and alkalis

**Ibn Sina** scientist living in the 900s in an area now in south Russia who described nerves and their uses in the body

**identical twins** twins formed when a fertilised egg splits into two identical halves, both of which develop into embryos

**image** view of an object produced by a lens or mirror

**imprinting** a rapid learning process that takes place in a young animal and establishes a behaviour pattern, such as recognition of parents

**infertile** unable to produce offspring

**inherited** passed on from a parent

**inherited variation** variation that has been passed on from a parent

**insects** arthropods with six legs

**instinctive behaviour pattern** a pattern of behaviour inherited from parents

**International Space Station (ISS)** research station in orbit around the Earth

**invertebrates** animals that don't have a backbone

**irreversible** a change in which new chemicals are made and it is impossible to get the original substances back again

**IVF (in vitro fertilisation)** when fertilisation is done by doctors outside the body, often in a glass dish

## J

**joule (J)** unit of energy

**Jupiter** a large outer planet made of gas and fifth in order from the Sun

## K

**keys** a way of classifying things into groups using their features

**kilogram** unit of mass

**kilojoule (kJ)** 1000 joules

**kilometre per hour** unit of speed

**kinetic energy** movement energy

## L

**lenses** devices made of shaped glass which focus light rays from objects to form an image

**light energy** given out by luminous objects which enables us to see things

**light microscope** a type of microscope which uses light to make images

**light, speed of** the distance light travels in one second (300 million metres per second)

**light year** the distance light travels in one year

**limescale** the everyday name for calcium carbonate

**Linnaeus, Carl** a Swedish scientist in the 1700s who invented the system of classifying plants

**liquid** a fluid with a fixed volume that takes the shape of its container

**lodestone** a natural magnet

**lubrication** a substance which reduces friction between surfaces when they rub together

## M

**Maglev train** train that floats, supported by a magnetic field

**magnet** something that attracts magnetic materials

**magnetic field** area around a magnet where there is a force on a magnetic material

**magnetic** material that is attracted to a magnet, e.g. iron, steel, nickel, cobalt

**magnetised** made into a magnet

**magnetism** a force which attracts certain metals, e.g. iron and steel

**magnify** make bigger

**mains electricity** electricity generated in power stations and available through power sockets in buildings

**male reproductive system** the organ system in males for producing offspring, consisting in humans of testes, sperm ducts and penis

**mammals** warm blooded vertebrates that have hair or fur and suckle their young

**Mars** a rocky inner planet fourth in order of planets from the Sun

**mass** the amount of matter in something

**mass of solution** the mass of a solution reached by adding together the masses of the solute and solvent

**material** made of matter

**matter** a substance that takes up space and has mass

**melt** turn from a solid into a liquid

**melting point** temperature at which a solid turns into a liquid

**menopause** stage when egg release and periods stop in a woman (at about 50 years old)

**menstrual cycle** monthly cycle in women of egg release followed by breakdown of the uterus lining

**Mercury** the rocky inner planet nearest to the Sun

**metal** one of a group of elements having certain similar properties, for example shiny, good conductor of heat and electricity

**methane** a gas which is a type of hydrocarbon (containing carbon and hydrogen atoms)

**metre per second** unit of speed

**microscope** a piece of apparatus that makes objects look bigger by making an enlarged image of them

**Milky Way** the galaxy containing our Sun and Solar System

**minerals** chemicals like calcium and iron that are needed to keep the body healthy

**mixture** containing two or more elements or compounds mixed together

**Moon** rocky body orbiting the Earth which is its natural satellite

**Moon, phases of** the parts of the Moon we see as it orbits the Earth

**moons** the natural satellites of planets

**moss** non-flowering plant with no true roots or leaves which reproduces by spores

**movement energy** energy that makes things move, also called kinetic energy

**muscle cells** special cells which make up muscles

**muscle fibres** muscle cells are arranged in fibres inside the muscle

**muscles** tissues made of cells that can contract and cause movement

## N

**natural gas** a fossil fuel which collects above oil deposits underground

**Neptune** a large outer planet made of gas and eighth in order from the Sun

**nerve** long, thin strand of nervous tissue that connects all parts of the body to the brain and central nervous system

**nervous system** the organ system consisting of brain, spinal cord, nerves and sense organs which sends messages around the body

**neutral** a substance with a pH of 7 which is neither acidic nor alkaline

**neutralise** to add acid or alkali to a solution to make a solution of pH 7

**neutralisation** the process of making a solution neutral

**newton** unit of weight

**Newton Isaac** English scientist in the 1600s who defined the law of gravity and invented the reflecting telescope

**newton meter** a device used to measure weight in newtons

**nicotine** an addictive drug found in tobacco

**night time** the period on one section of the Earth when it is facing away from the Sun

**nitric acid** a strong acid

**non-metals** elements that are not metals

**non-renewable (energy)** energy source that cannot be used again and will run out eventually

**normal brightness** standard brightness of a single bulb lit by a single cell

**northern hemisphere** the half of the Earth between the equator and the North pole

**nucleus** the part of a (living) cell which controls all its activities

**nutrient** useful substance derived from food

## O

**oil** a fossil fuel formed from sea creatures over millions of years

**orbit** the path taken by one body in space around another (e.g. Earth around the Sun)

**omnivore** animal with a mixed diet of animals and plants

**organ** group of tissues working together to do something useful

**organ system** group of organs working together to carry out different life processes

**ovary** female organ where egg cells develop

**oviduct** tube leading from ovary to uterus (womb)

**oxygen** a non-metal element that exists as a gas in the atmosphere and is needed for life and burning reactions

## P

**parallel circuit** electric circuit in which there are two or more paths for an electric current

**particle theory** theory to explain how matter behaves

**particles** tiny pieces of matter

**penis** male organ involved in reproduction – deposits sperm into the vagina of the female

**periods** the everyday word for the part of the menstrual cycle when the lining of the uterus comes away as blood

**petrol** a hydrocarbon fuel (containing hydrogen and carbon) which comes from crude oil

**pest** organism that damages other organisms or spreads disease

**pH** a way of measuring how acid or alkali a substance is

**photosynthesis** how green plants make their own food

**pH scale** the range of levels of acidity and alkalinity

**phytoplankton** microscopic plants found in water

**placenta** organ that links the blood supply of a growing fetus with the mother for transfer of nutrients

**planet** any large body that orbits a sun in a solar system

**plant cell** a cell which forms part of a plant and has a cell wall and often chloroplasts

**plants** organisms that make their own food

**Pluto** used to be seen as the ninth and last planet from the Sun, now called a dwarf planet together with others of the same size beyond it

**polar orbit** orbit that travels around the North and South poles of the Earth

**poles, of Earth** the north and south points of the Earth connected by its axis of tilt

**poles, of magnet** the opposite and most strongly attractive parts of a magnet

**potential energy** stored energy

**power station** place where fuel is burned to produce electricity

**predator** animal that catches and kills other animals for food

**predict** say what you think will happen

**pregnancy** the period of time taken for mammal offspring to grow and develop inside their mother's body

**premature** before the expected time – in human babies it means born before 38 weeks

**prey** animal that is hunted and killed by other animals

**products** new substances made in a chemical reaction

**property** a characteristic of a substance

**Proxima Centauri** the nearest star to our Sun

**puberty** age when males and females become sexually mature

**pure water** water produced by distillation which has no solutes dissolved in it

## R

**range** the values a measurement can have

**react** undergo a chemical reaction

**reactants** substances that react together in a chemical reaction

**red blood cells** cells in the blood that are specialised to carry oxygen

**reflecting telescopes** powerful telescopes which use concave mirrors as well as lenses

**refracting telescopes** telescopes which use two lenses

**renewable** (energy) energy source that can be replaced and won't run out

**repel** push away

**reproductive system** organ system for producing offspring (also see female and male reproductive systems)

**reptiles** vertebrates with scaly skin that lay waterproof eggs

**reversible** a change in which you can get the original substances back again

**s**

**salts** chemicals formed by neutralisation reactions

**satellite** any body that orbits another

**satellite navigation (satnav)** navigation system used by ships and motorists which is linked to satellites

**saturated solution** a solution in which no more solute can dissolve

**Saturn** a large outer planet made of gas and sixth in order from the Sun

**scanning electron microscope (SEM)** a type of electron microscope

**scatter graph** graph that shows all the values in a set of measurements

**seasons** changes in the climate during the year as the Earth moves around its orbit

**section** very thin slice

**seed** produced by flowering plants and conifers – grows into a new plant

**semen** fluid containing sperm and nutrients

**sense** be aware of something

**sense organs** organs that are specialised to respond to stimuli such as light, touch, taste, smell and sounds

**series circuit** components are joined in a single loop

**sex** the act between male and female organisms which allows their sex cells to meet and form new organisms

**sex hormones** hormones that cause the changes which happen at puberty

**signal** something in the environment that can trigger an instinctive behaviour pattern

**skeleto-muscular system** the organ system which brings about movement of the body, consisting of the bones in the skeleton and muscles

**skeleton** system of bones inside the body

**sodium hydroxide** a strong alkali

**solar energy** energy from the Sun which can be used directly to make electricity or for heating

**Solar System** our Sun and the planets and other bodies that are in orbit around it

**solenoid** a coil of wire

**solid** a state of matter with particles arranged in a regular pattern

**soluble** able to dissolve

**solute** a substance that dissolves in a solvent to make a solution

**solution** a mixture of solute and solvent

**solvent** liquid part of a solution

**sound** a form of energy given out by something that makes a noise

**southern hemisphere** the half of the Earth between the equator and the South pole

**Space** the area outside Earth's atmosphere and surrounding all bodies in the Universe such as stars, planets, etc.

**Spallanzani, Lazarro** a scientist in the 1700s who proved that sperm and egg had to meet for fertilisation to happen

**specialise** have special features to perform a particular function, e.g. specialised cells

**specialised cells** cells which have special features to perform particular functions

**species** group of organisms that can interbreed and produce fertile offspring

**speed** the distance travelled in a given time

**speed of light** the distance light travels in one second (300 million metres per second)

**sperm cell** male sex cell

**sperm duct** tube which carries sperm from the testes to the penis

**spore** produced by ferns and mosses – can germinate and grow into a new plant

**springs** metal wound into spirals which can store elastic potential energy

**stars** bodies in space which give out their own light

**states of matter** solid, liquid and gas

**stored energy** energy which is stored, for example in food, fuel, batteries, elastic materials and objects raised high up

**streamlined** shaped to reduce resistance to motion

**stretching** when elastic materials such as springs are pulled outwards or downwards

**strong acid** acid with a pH of 3 or less

**strong alkali** alkali with a pH of 11 or more

**sulfuric acid** a strong acid

**Sun** star at the centre of our Solar System

**sunlight** light from the Sun

**sustainable** (energy use) energy use that can be supported by renewable sources

**symbol** a sign which represents something (also see chemical symbols, circuit symbols and hazard symbols)

## T

**telescopes** devices made with lenses which allow distant objects to be seen clearly

**TEM** type of electron microscope

**testis** organ where sperm are made

**theory** an idea to explain something

**thermal energy** heat energy

**thrust** force from an engine or rocket

**tidal energy** energy from water in tides which can be used to generate electricity

**tissue** group of similar cells

**transfer** (of energy) moving energy from one place to another, e.g. from electricity to light

**triplets** three babies that have developed in the womb at the same time

**twins** two babies that have developed in the womb at the same time

## U

**unbalanced forces** when the opposite forces on an object are unequal so the object moves

**universal indicator (UI)** substance that changes colour to show the pH of the solution It's mixed with

**Universe** everything that exists around us and beyond

**upthrust** upward force

**Uranus** a large outer planet made of gas and seventh in order from the Sun

**uterus (womb)** place where a baby develops and grows

## V

**vacuole** large 'bag' inside a cell containing sap

**vagina** tube from the uterus (womb) to the outside through which a baby is born

**variation** differences between individuals

**Venus** a rocky inner planet second in order of planets from the Sun

**vertebrates** animals that have a backbone

**volt** unit of measurement of voltage

**Volta, Alessandro** Italian scientist who investigated electric current

**voltage** measure of electrical energy

**voltaic pile** predecessor to today's alkaline batteries, made by Volta

**voltmeter** device for measuring voltage

## W

**water** clear liquid which forms a large part of all living matter

**wave energy** energy from water in waves which can be used to generate electricity

**weak acid** acid with a pH of 5 or 6

**weak alkali** alkali with a pH of 8 or 9

**weight** the force of gravity on the mass of an object

**wind energy** energy from wind which can be used to generate electricity

**womb** another word for uterus in the female reproduction system

**word equation** representation of a chemical reaction in words

# Index